WOMEN AND TEACHING

WOMEN AND TEACHING

Global Perspectives on the Feminization of a Profession

Regina Cortina
and
Sonsoles San Román

WOMEN AND TEACHING

© Regina Cortina and Sonsoles San Román, 2006.

Softcover reprint of the hardcover 1st edition 2006 978-1-4039-7309-2

First published in 2006 by
PALGRAVE MACMILLAN™
175 Fifth Avenue, New York, N.Y. 10010 and
Houndmills, Basingstoke, Hampshire, England RG21 6XS
Companies and representatives throughout the world.

PALGRAVE MACMILLAN is the global academic imprint of the Palgrave Macmillan division of St. Martin's Press, LLC and of Palgrave Macmillan Ltd. Macmillan® is a registered trademark in the United States, United Kingdom and other countries. Palgrave is a registered trademark in the European Union and other countries.

ISBN 978-1-349-53434-0   ISBN 978-1-4039-8437-1 (eBook)
DOI 10.1057/9781403984371

Library of Congress Cataloging-in-Publication Data

Women and teaching : global perspectives on the feminization of a profession / edited by Regina Cortina and Sonsoles San Román.
    p. cm.
Includes bibliographical references and index.

1. Women teachers—History. 2. Feminism and education—History.
I. Cortina, Regina. II. San Román, Sonsoles.

LB2837.W655 2005
371.10082—dc22   2005053508

A catalogue record for this book is available from the British Library.

Design by Newgen Imaging Systems (P) Ltd., Chennai, India.

First edition: April 2006

10 9 8 7 6 5 4 3 2 1

Transferred to digital printing in 2007.

CONTENTS

List of Figures

List of Tables

Acknowledgments

We are grateful to the *Instituto de la Mujer, Ministerio de Asuntos Sociales de España* (Women's Institute, Ministry of Social Affairs of Spain) for providing the funding to translate some of the chapters included in this book. Their support was especially timely since this is the first collection of essays that focuses on the feminization of teaching from a global perspective. We thank the *Instituto de la Mujer* for encouraging this new line of interdisciplinary research connecting Latin America, Europe, and North America in a comparative research enterprise.

We would also like to express our gratitude to the Colegio de San Luis Potosí in Mexico, which sponsored the First International Congress on the Feminization of Teaching in February 2001, where this book had its origins. Researchers who have devoted years to studying the subject had the opportunity to meet during this conference. We have compiled several of the papers initially presented there in this book, supplemented by additional selections.

This book could never have been completed without the talented and careful work of our translator Margaret Carson, who has brought the reader a finely crafted translation of the original Spanish. We are also grateful for the editing done on the Spanish originals by Rafael Morales Barba, Professor in The Universidad Autónoma de Madrid.

We also want to acknowledge the support that Regina Cortina received from the Steinhardt School of Education at New York University in the preparation of this book. In editing and revising the manuscript, we owe special thanks to Lucía Cárdenas. We would also like to thank the School of Education at University of North Carolina-Chapel Hill for their support as well. A special thanks to Janet Lopez for the final revisions on the manuscript. Finally, we want to acknowledge the support that Sonsoles San Román received from the Universidad Autónoma de Madrid to travel to New York City to work on the conceptualization of this book in the initial writing stages.

Regina Cortina
Sonsoles San Román

Introduction: Women and Teaching—Global Perspectives on the Feminization of a Profession

Regina Cortina and Sonsoles San Román

This book is intended to fill a gap in the historical and social research into contemporary educational systems. Our purpose is to use the frame of gender, with all of its complexities, to analyze the role of the woman teacher from a comparative perspective. The chapters that follow will be of great use to all those who are interested in removing the historic veil over the feminization of teaching, making it possible to discover the diversity and plurality of the social contexts in which a feminized profession has been produced.

Pursuing this line of inquiry, we will examine from a comparative and global perspective the historic, political, social, and religious contexts which give shape to the educational structure in countries that are considered in this volume. Throughout these pages, we will attempt to show the degree to which the presence of women as a majority of teachers at the early levels of the educational system is rooted in certain historical, educational, economic, and political factors that still prevail today. One of our key objectives is to explain the forces that shaped the process of feminization in teaching.

The existing studies on this subject are, for the most part, historical investigations regarding the process of feminization in Britain,[1] Scotland,[2] Canada,[3] Australia,[4] New Zealand, and the United States,[5] countries in which the majority presence of women in the teaching field manifested itself almost a century earlier than in most of the countries for which case studies are presented in this volume. In 2003, for the first time a French journal published a volume comparing research on gender in the history of teaching in English-speaking countries with the histories of teaching in European countries.[6]

In English-speaking countries, a predominance of women in both urban and rural educational systems was already evident by the second half of the nineteenth century. In the case of urban schools, social historians in the United States have concluded that the factor that favored women's entry into this occupation was the organization of schools by grades; women teachers were thus concentrated in the early grades and men teachers in the higher grades and in school administration.[7] But this thesis cannot be easily adapted to the expansion of rural schools, because in some cases these schools were organized as one-room schools, while in others they were organized by grades.[8]

In the academic writings about the feminization of teaching, scholars have asked why men abandoned the classroom when women began to arrive. Among the explanations that have been developed, the ones pointing to changes in the bureaucratic organization of schools should be emphasized. When the school year became longer, there was a corresponding increase in professional requirements for teachers; but surprisingly, the salary remained the same. Confronted by this situation and able to pursue other job opportunities, men teachers abandoned the schoolroom at an accelerated rate.[9]

Through an extensive empirical and comparative study of the historical record of feminization in the different regions of the United States, Perlmann and Margo[10] show that feminization extended to rural zones not only because of the bureaucratic organization of the schools—since many of them were one-room schools—or the fact that women were paid less than men, but also because they were considered to be more maternal and adept at working with young children.[11] In contrast to studies that explain the massive process of feminization from a purely sociological or economic perspective, these authors find that the salary of women teachers did not increase—even despite a high demand—because traditional gender patterns restricted the participation of women in other occupations in the labor market. Perlmann and Margo conclude that the social transformations that parallel the evolution of school organization and cultural changes were "inextricably tied up" with the gendered character of teaching and the view that it was primarily a woman's occupation.

Since the mid-1990s, several anthologies have been published by social historians, educators, and feminist scholars focusing on the cultural construction of teaching from a gender perspective. These works seek to explain how gender shapes the social and school realities of women teachers, while also taking into account how these realities are conditioned by women's own social characteristics, such as race, social class, marital status, age, and so on.[12] The division of pedagogical and administrative responsibilities by gender in schools presents an area of investigation in which researchers have asked how women might succeed in overcoming gender barriers to achieve positions of leadership in public school employment.[13] Within all these published works, many of them based on ethnography and interviews, we hear the dissenting voices of women, organized during some periods of history while at other times audible only through the voices and writings of women reformers. But in spite of their efforts, even today the issue of gender is not addressed in many programs that prepare new teachers for the profession. Nor is the question of how the cultural construction of teaching as a gendered occupation influences the identity of teachers as they enter the classroom.[14]

The stratification of employment by gender within education, or the process of feminization, occurred because of two parallel trends. While on the one hand a transformation of social values took place—a situation that opened new spaces for women in the labor market—on the other hand conventional gender roles remained intact. The supervision and control of women's work remained in the hands of men, as principals and supervisors, while the work of women teachers was limited to the space of the classroom.

In the cases described within the studies collected in this volume, we can observe that the feminization of teaching was not solely due to abandonment by men and an influx of women, since in many countries the teaching profession was feminized since its inception. These case studies underline the importance of understanding political power and the political discourse that shaped to the development of public educational systems and encouraged the entrance of women into the teaching profession.

How, when, where, and why did women teachers begin to enter schools on such a large scale? In contrast to English-speaking countries, Catholic countries in Europe, Latin America, and the Caribbean saw the entrance of women into schools specializing in the training of teachers during the last two decades of the nineteenth century. This incorporation coincided not only with the institutionalization of teaching on a national scale, but also with the resulting secularization and centralization of educational systems by the State. It would be only later, as a result of policies to expand public educational systems—a phenomenon that occurred in most countries in the first half of the twentieth century, which coincided with the influx of women into teaching—that the feminization of the profession was produced.

The accelerated growth of universal education entailed a transformation of the political discourse to support the growth of the economy and deal with the demands that arose once national systems of education were established. Politicians, legislators and intellectuals were responsible for creating an environment favorable to the incorporation of women into teaching, and of convincing these women of their innate calling as teachers.

The chapters of this volume explore how the demand caused by the opening of schools for girls, together with the resulting political and social needs, led to the hiring of educational personnel at low cost. Given this situation, the woman teacher became an economical option in the eyes of the authorities in charge of hiring them. With such a decrease in cost, these authorities obtained an important economic benefit because the salary of women teachers (legislated in only a few countries by law) was, in fact, lower. As the field expanded in the first half of the twentieth century, men gradually abandoned teaching because it did not compensate them adequately. In this way, and as a result of the same economic development, there were new possibilities to enter the job market for men or new possibilities for public service and political action. Such was the case in Mexico, where men left the schoolroom to become school administrators or participate in union and political parties, while women for the first time had the opportunity to enter the labor market with a skilled and remunerated job.

The educational hierarchy, which is differentiated by grades, opened spaces to accommodate women teachers at the bottom of the educational pyramid, that is, in the early grades, which have the least status and lowest pay. While in some countries laws were passed mandating that preschool be taught by women, in others this requirement also included the early grades of elementary school. In certain countries, this requirement made it necessary to open teacher training schools specializing in preschool and primary education only for women, which clearly contributed to the feminization of this educational level. Along with that pattern, we find that in many countries, teacher training schools for preschool and elementary grades, and all the positions for principals in these schools, have been reserved for women, a tendency documented on several occasions by international organizations.[15]

The historical and comparative research that studies the process of feminization in teaching explains, in part, the dynamics that set in motion women's entrance into the field of education. In countries such as Costa Rica, Spain, or Mexico, wives, mothers, and daughters were enlisted to help men teachers, a highly economical solution to respond to demands placed by central policies that decreed universal schooling. These women teachers entered teaching by way of kinship without receiving any remuneration for their work, while at the same time, in the absence of teacher training schools for women, men teachers gave them basic instruction in reading and writing. The social image of the profession was overwhelmed by its female composition. It was widely assumed that the presence of a woman teacher in the schoolroom was required as a natural complement to the man teacher in order to contribute a maternal quality. Denying women teachers the professional level required of men teachers, schools became a convenient institution to socialize young children in traditional gender roles. Years later, these gender differences were still maintained within the teacher training schools themselves. While the "practical" schools trained women for primary school education, as in the case of the Dominican Republic and Brazil, the schools for upper-grade teachers were set aside for men.[16]

The social function fulfilled at home by women, the agent to whom the care for young children was delegated, supported their absorption in schools. Thus the role that women traditionally occupied in the domestic space began to extend itself into the public sphere. That is to say, the demand for women did not occur because of requirements for professionalization; on the contrary, their entrance into teaching should be understood as a result of the glorification of the so-called feminine nature that made a woman a suitable candidate to be put in charge of young children in her role as social mother. The presence of women was required once their maternal qualities were exalted and they were placed on the pedestal of "substitute mothers" or "a mother made conscious" (a term coined by Friedrich Froebel),[17] or even more preferably, widows, who were delegated a space of power for the first time to carry out a skilled job in the school.

The establishment of coeducation in schools marked a fundamental transition for women in teaching. For moral and religious reasons, women had not been previously allowed to teach both girls and boys, which meant that school authorities had to make a double outlay, hiring men teachers for boys' schools and women teachers for girls' schools. But after coeducation was customary, the woman teacher no longer worked strictly in girls' schools. She could now also teach boys and adolescents of both sexes. Keeping in mind that coeducation occurred due to political pressures for universal compulsory education— and was not obligatory in all countries—it becomes possible to understand why this process took place primarily in public schools. Private and religious schools resisted, an opposition led by the Catholic

Church and reinforced by traditional families who understood that it was necessary to maintain a differentiated socialization within the classroom and the school to keep gender roles intact.[18]

It is necessary to make a distinction between coeducational and mixed schools. While the system of coeducation was established during modernization in the early twentieth century as a political requirement—which gives it a positive cast—mixed schools were opened due to economic factors in municipalities lacking the resources to offer separate schools for girls and boys. Significantly, in some countries two separate entrances were used in the same building, and there were even two sessions in order to offer segregated education for the two sexes despite economic scarcity. An interesting and more recent case is that of Belgium, where the practice of educating girls and boys in different schools also gave rise to separate teacher training colleges for men and women. It was not until 1983 that coeducation in Belgium was mandated, due to pressure placed by the European Union in light of demands for equal admissions for men and women in vocational and professional schools.

Feminization and Employment of Women

Certain relationships between feminization and professionalization must be underlined. One of the conclusions of the studies presented in this volume on the phenomenon of feminization during the twentieth century points toward the lack of professionalization within the teaching field. This profession is associated with a low social status, low salary, lack of authority and discipline in the character of the woman teacher, loss of accumulated experience as married women left teaching, or the mentality of the inspectors, who demanded that the female teaching staff display maternal qualities, such as patience, a sweet disposition, and love for young children, to the detriment of their professionalism.

In the specific case of women teachers, the studies continue to highlight the asymmetries that exist between men and women in their organizational participation within schools. The incorporation of women into the teaching field was due, in great part, to the belief that their presence would offer young children the "innate" feminine qualities that men lack. As a consequence, women are still considered today as the bearers of these qualities, which are related to their "feminine condition." Such a social demand limits the requirements of professionalism and at the same time fills the woman teacher with anxiety as she faces pressure from families and authorities to carry out a function that is more maternal than professional.

When convinced that their maternal qualities make them irreplaceable agents to be put in charge of young children, women have been the essential element through which the State has achieved the expansion of public education over the past century. The price of achieving such a goal has been detrimental to the levels of professionalism extended to women, as these demands for maternal-like behavior do not advance the woman teacher's technical-pedagogical knowledge. Moreover, this requirement limits women teachers to the preschool and primary school grades, impeding their access to secondary or higher levels of education.

A correlation also exists between the admission requirement for teachers and the need of the State to extend universal schooling across the nation. To satisfy this demand, the State utilized women, a cheaper source of labor entering the field in massive numbers once the entrance requirement—years of schooling, professional requirements, and wages—were lowered. For this reason it can be observed that levels of feminization have been at their highest during the historic moments when more teachers are needed. During these periods, women teachers are used to fill the required spaces. The achievements made by the State in expanding its educational services have been obtained thanks to the definition of teaching as feminine, which permits the recruitment of women at relatively low cost in response to the demands of education.

The role of the Catholic Church, the pillar upon which social structures and traditional gender roles rest in Spain, Belgium, Latin America, and the Caribbean, must be emphasized in discussing the process of feminization. Its message has helped to preserve a mentality that accepts women only in professional jobs that are compatible with motherhood. Thus the division of functions within schools is guaranteed, despite the efforts undertaken by the State to secure coeducation and the equal insertion of women in the job market.

The studies included in this book offer keys to understanding how women teachers experience a feminized profession; how this experience affects the development of school culture; how women teachers educate girls and boys with a view toward their future job participation; how women teachers participate in unions; how gender patterns are incorporated within schools; the way in which teachers assume their profession; and the extent to which this gender identity affects the teaching profession.

The educational policies sponsored by international development agencies in the last decade are also an important factor to consider. International agencies select policies that devalue practical and pedagogical knowledge, profoundly damaging the contribution of women as educational professionals and detracting from their job competence. These agencies have deemed that professional work in a preschool is at the same level as the care that a neighbor provides for children while their mothers and fathers work.[19] The new educational reforms continue with this policy of submission and docility, and utilize new means to continue the deprofessionalization of women's work in education by proposing that a neighbor or a young adult who has taken quick courses in pedagogical training[20] can be put in charge of the youngest or the poorest children.

When women are concentrated in the lower grades of schooling without access to administrative and leadership roles, the feminization of teaching presents the danger of destroying their aspirations to take part in the political process regarding educational policies that affect them. Without such participation, the mandates from political agencies and governments are thrust upon teachers without warning and make them change their activities in school from one day to the next, creating confusion and stifling their professional expectations. To the extent that administration and policy-making become arenas closed to their participation, a division is created between the woman teacher and the management and direction of schooling. As a result, in most countries, women are the ones who teach, while men occupy positions in administration, planning, and in the areas that promote and strategize educational reforms.

In certain historical moments of social change during a process of political transition, political militancy and an increasing aspiration to occupy administrative positions within schools can be observed among women teachers. In the case of Spain,[21] the teachers from the generation of 1968—the transition period before the consolidation of parliamentary democracy—went on strike and demonstrated in favor of their political rights. They literally "wore the pants," a symbol of masculinity, as they entered their classes and broke with the submissive image required by the inspector. Such behavior, which began to contradict the conformist image the authorities demanded of women teachers, helped to modify the political ideology from that space of power represented by the classroom. In the case of Mexico, women teachers in the 1930s organized and fought for their political rights, promoted coeducation in public schools, and achieved equal labor rights as government workers.[22]

Consequences of Feminization on the Teaching Staff and Students

One of the conclusions from studies included in this book refers to how the feminization of teaching affects the culture of schools. In some cases the preponderance of women in schools creates a feminized atmosphere that favors working in groups and fosters a team spirit between the teachers. But in most cases this climate leads to an isolation that reflects women's distance from spaces of power. When the professional identity reinforced by educational institutions does not support aspirations among women teachers for professional and political influence, it is more difficult to create a school culture that fosters participation and collaboration.

In addition, the feminized atmosphere infiltrates schools and gives rise to gender patterns that also affect men. To the extent that women's work is identified with school at the early childhood level, men are not considered socially and emotionally suitable to carry out this work. They exclude themselves from teaching at preschool and primary levels and channel their professional aspirations in other directions. Beyond the threshold of feminized work in these early grades, an atmosphere charged with masculine symbols and patterns opens up in the upper grades. These patterns are reflected once again in the structure of the job market. The attainment of equal opportunities can be achieved only by eliminating the rigid gender hierarchies that are in the very air that young children breathe inside schools, or by placing women and men teachers at an equal rank.

Despite the predominance of women teachers in schools, the standards of behavior to be followed are set by the senior administration, the majority of whom are men. For example, we attended an event at a public school in New York City to listen to a talk by a well-known researcher (a woman) about Mexican immigration to the borough of Queens. The educational center intended to inaugurate the school year with this event and the entire teaching staff was assembled. To our surprise, the principal burst into the auditorium and went up to the platform, there to keep more than two hundred women teachers, the invited guest included, seated on the benches in the room. He opened his speech by ordering the teachers to begin classes by writing the objectives of that day's lesson on the blackboard, warning them that he himself would walk the hallways to monitor their compliance with his orders. After that he referred to the studies of Howard Gardner[23] about multiple intelligences to demonstrate his superior knowledge to an auditorium of women whom he appeared to regard as relatively ignorant. Finally, he offered to give them a formal lesson explaining how to maintain discipline in the classroom. The strangest part was that the middle-aged women listening to him had many years of teaching experience and a high degree of professionalism. Nevertheless, the principal wielded his masculine presence, accompanied by a tone of authority and power, and seemed to be strengthened by the submissive audience, who without making a sound endured such treatment in courteous silence. In this way, much like the Foucault's microphysics of power,[24] he maintained for an hour and a half a discourse of masculinity, power, and authority. Once it was the researcher's turn to speak, the principal made a point of interrupting her several times during her presentation.

The feminization of teaching, which without a doubt influences the socialization of children, has an effect on the equality of opportunities between the sexes, above all in the transition from primary to secondary school—a process that takes place between the ages of eleven and twelve, at which time the children no longer have women teachers in subjects like mathematics and natural sciences.[25]

Why do girls receive better grades in certain subjects? While in English-speaking countries there are abundant studies examining this question,[26] the absence of studies in Spanish-speaking countries shows the degree to which researchers have not assigned importance to studying the effect that the feminization of teaching has on the performance of children and the professional expectations held for them.[27] In this sense, we concur with those studies that have examined how as a result of the feminization of teaching the performance of girls and boys is affected due to possible gender identification. This situation leads girls to receive higher grades in areas such as languages, humanities, music or art, while at the same time having negative repercussions on the configuration of their professional expectations. We believe that one way to understand why women opt for a profession that is compatible with motherhood is to investigate the school itself. The levels of feminization of teaching, although they vary from country to country, remain stable in those countries once the profession is feminized, except in times of severe economic dislocation when more men return to the profession until they can again secure other opportunities.

Surprisingly, the success experienced by girls in school does not correspond to their workforce participation or to the managerial spaces they occupy outside school, where they collide with a social atmosphere that relegates them to subordinate positions.[28] An unemployment line or "labor queue"[29] determines that women are unemployed at higher rates than men even when their academic preparation is higher, and they more frequently must perform jobs that they find much below their educational level. Thus, while school presents itself as the "promised land" for women,[30] the labor market moves them toward spaces in the lower ranks, which are traditionally considered as feminine and compatible with their maternal role.

In university classrooms there is, more and more, a greater presence of women in majors traditionally considered masculine, despite the fact that the labor market may not necessarily reflect this tendency. Only in the case of the so-called semi-professions[31]—such as teaching, nursing, or social work, among others—is there an equal percentage of women students who major in these fields and women professionals who work in them. Today, when there is presumably no impediment that bars a woman from choosing any profession she wishes, the fact that women continue to opt for these fields points to a significant degree of self-exclusion.

All evidence seems to indicate that the responsibilities and functions that the woman assumes in the home, arising from her role of mother of the family, leads her to understand motherhood not only from a biological point of view, but also as an exclusive sphere of feminine social identity. And that is how women over time develop abilities that allow them to enter certain spaces of the labor market with ease. From this perspective, it is possible to understand why women teachers themselves strongly defend women's qualities in carrying out their jobs. But in this respect we would also warn that by prioritizing the value of love toward young children, the ascending line of professionalization in the teaching field is put in danger. In effect, the feminized space that is experienced in the school creates a climate that favors her incorporation at the bottom of the educational system. As a result, the relations between early childhood and motherhood still condition the social function of the women teachers in the public school.

DESCRIPTION OF CHAPTERS

For the purpose of expanding the contributions of historical and comparative studies on the causes that produced the feminization of teaching and maintaining it to the present day, we have selected the following case studies about specific countries. The different inquiries that comprise this book are divided into three sections.

In the first section, the way in which women teachers experience and construe their feminized profession is explained. To study the meaning of the term *vocation* and understand how women teachers from two historical–generational spaces have assumed the feminization of schools, Sonsoles San Román employs a qualitative focus to interpret key aspects of professional identity among women teachers in Spain during the transition toward democracy. In the first case that examines the generation of women teachers from the conservative culture of the middle years of Franco's regime (1950–1960), we observe an assumed vocation linked to religious sentiments that clearly agreed with the ideology of national Catholicism and that demanded from women a spirit of sacrifice and devotion, accompanied by the absence of expectations of economic reward. In the second case, women teachers from the post-1968 period denounced the "vocational" nature of their profession by pointing out the lack of possibilities for women from rural villages to enter other professions. The comparative focus of this author's sociological study shows the level of power held by religion over the labor distribution of men and women in Spain.

The feminized culture of the school, particularly its effects on the professional development of teachers, is the focus of the three case studies that follow. Sandra Acker's previous research emphasizes the degree to which the high percentage of women concentrated in the lower years of the educational system are insufficient, in and of themselves, to explain the kind of feminized patterns that exist in school culture. By thus distancing herself from what is obvious from a purely quantitative vantage point, Acker shows the extent to which gender models—which affect both men and women teachers— emerge from the same societal expectations that lead us to presuppose that the individuals who work with young children should display a set of characteristics appropriate to what has traditionally been associated with the qualities demanded of women: patience, gentleness, flexibility, and the like.[32]

In chapter 2 presented in this volume, Elisabeth Richards and Sandra Acker analyze two case studies, exploring the reasons for what appear to be contradictory findings in the two settings. Sandra Acker's study took place in a primary school in England, while Elisabeth Richards interviewed core French elementary school teachers in Ontario, Canada. The authors were interested in the degree to which the environment inside these institutions, whose teaching staff is predominantly female, encourages caring, collegiality, and cooperation among the teachers. One of their findings is that while British teachers enjoyed a close, collaborative culture, Canadian teachers did not have the same experience. These teachers taught French to English-speaking children, and in the process moved from classroom to classroom, and often from school to school. Thus they were not in a position to be incorporated into a close teacher culture, and instead they were marginalized.

Following the same line of research, but in Argentina, Graciela Morgade shows how women's incorporation into teaching represented both an area of autonomy that allowed them to leave the home and undertake remunerated work, as well as a space where they were subjected to subordination and kept at the base of the pyramid of the school system, where they found only limited possibilities for professional development. Morgade examines how the feminized culture experienced in the classroom did not foster self-esteem among women teachers. This trend creates a mentality that favors incorporation into the lowest paid, least prestigious jobs, and leads women to the lowest levels of the system, thus reinforcing their self-exclusion from positions with greater power, higher prestige and better salaries, which are reserved for men.

The second section of the book includes several interpretations of the feminization of the teaching profession from different disciplinary perspectives. To explain how the image of the woman teacher is created through political discourse, and to explain the gender ideology that assigns certain hierarchical spaces to men while reserving others for women in both the profession itself and in the teachers' union, Regina Cortina examines writings and political discourse that define the social identity of women teachers and the work they do on the basis of a "vocation to serve."

During the years when the Mexican nation was consolidated and, as a result, public education expanded, José Vasconcelos, the Minister of Education, took advantage of Gabriela Mistral's visit to Mexico to promote the participation of women in teaching as a dignified, useful profession, thus legitimizing their participation in a salaried occupation. Cortina affirms that Mistral's legacy still endures in Mexico, where the conception of teaching as "women's work" has never been modified and, consequently, the professionalization of teaching has progressed very little.

In addition, Mexico's political economy has had a devastating effect on the teaching profession, which has lost its status and position as a middle-class occupation. As a result, the praise routinely directed at the work of both men and women teachers is merely rhetorical, given the lack of support needed to strengthen their functions in the classroom and to improve their opportunities for professional development and increased salaries. Though the first feminist teachers at the turn of the twentieth century fought to achieve coeducation and better working conditions for women in teaching, in the twenty-first century we now find a profession that is impoverished, lacking in opportunities for professional advancement.

In the case of Brazil, Fúlvia Rosemberg emphasizes the impact of the expansion of educational reforms on men's and women's education. She then inquires into how the dynamics of femaleness and maleness are preserved through schools. After looking into factors such as gender inequality between white and black women and the average years of schooling in both groups, the author analyzes the divergent specializations of men and women in disciplines and areas of knowledge, as well as their participation in different levels of the educational system as teachers. Her research supports the hypothesis that their marked segregation in the labor market makes it difficult to change the professional formation that men and women receive. The distinct trajectories of the two groups at different levels of the educational system cannot be explained only by focusing on salary and the type of educational training.

Rosemberg concludes that the trajectories followed by men and women teachers can be explained by considering the age of the students, for while she found that women teach mostly children, adults tend to be taught by men. For this reason, most women teachers have only a high school education, while men have advanced to college levels. However, differences in the level of education are not sufficient to explain salary differences, because we find disparities at each level of instruction as well. Apparently, there exists an "invisible ceiling" that impedes women from gaining promotions to careers that are socially considered as 'male,' while an "invisible escalator" benefits men in professions defined as "female." Rosemberg's objective is to widen the agenda of research in order to identify tendencies in the interaction of gender, class, race, and age in schools, and their repercussions for the future prospects of the students in the labor market.

In an effort to comprehend the process of feminization beyond percentages and statistical tables and using the case of Belgium, Marc Depaepe, Hilde Lauwers, and Frank Simon propose a series of hypotheses that address an important gap in the research. Their quantitative historical analysis shows a rise of inequalities and an increasing vertical segregation of the profession—the presence of men in authority and administrative positions and the concentration of women in classrooms. Not only does the gender segregation increase, but the jobs of men and women become more and more distinct. While men pursue managerial and policy-making careers, women are concentrated in teaching roles, and their percentage as part-time workers also rises, which de facto limits their possibilities for promotion. The authors argue that to the extent that the level of feminization increases during periods of modernity, it is related to the transmission of gender patterns through education. The traits of the teaching profession as institutionalized in society coincide with the feminine characteristics demanded of the women teacher.

The chapters in the final section of the book attempt to understand the factors that cause the feminization of teaching from a historical, social, political, and economic standpoint. Using an economic perspective, Iván Molina analyzes the Costa Rican case, where the State searched for an economical solution to satisfy the demand for teaching personnel and thus deal with the growth that schools experienced after the reforms of 1886. The State, putting demands for professionalism to one side, decided instead to look for an economically feasible solution, and it was thus by virtue of their low-wage status that women gained access to the field of education. At the same time, the expansion of the urban economy offered better-educated men options to improve their positions at the level of primary education, a process that benefited from the differences in public perceptions about the type of work that men and women could perform. As a consequence of this situation, men controlled the best jobs. Molina also identifies kinship as another factor that explains the process of feminization. Because of the low salaries men teachers received, they invited their wives, mothers, and daughters to participate in teaching, and these allies became key to the feminization of the teaching profession.

In the case of the Dominican Republic, Juan Alfonseca explores how society and the curriculum contributed to the historical causes and politics that triggered the feminization process. His study focuses on the period during the nineteenth century when the schools switched from being an institution defined by the presence of men to one in which women became the main actors in teaching. His findings point to the differences between urban and rural zones and indicate that the increase in the number of women teachers in urban areas was closely related to the expansion of the middle class, while in rural areas the cause of this phenomenon was the abandonment and desertion by men of the teaching field.

Mexico is examined once again, but this time in light of the social figure of the woman teacher in Mexico in the late nineteenth century through letters they sent to President Porfirio Díaz. Luz Elena Galván uses these documentary sources to describe the ethos of women teachers in that era: their anxieties, their successes, their weak, sickly bodies, their hairstyles and fashions, and their condition of widow-hood and poverty. The contrast between women teachers of humble origins, who in Mexico, at least, embody the vast majority of teachers, and the achievements of a small number whose access to culture and education favors them, allow Luz Elena Galván to affirm that the dif-ferences among women teachers are determined by their unequal access to culture.

Further Issues for Discussion and Additional Research

We wish to conclude this introduction by encouraging other researchers to continue studying the social construction of women teachers in different socio-historical and political periods in each nation. The chapters included in this book point to interesting trends, showing how the feminization of teaching affects women's roles and employment in education. But the ways in which gender interacts with other social characteristics of women such as race, class, and socioeconomic status can only be illuminated through detailed social histories that include the historic, political, social, and religious con-texts that shape the participation of women teachers in the educa-tional profession.

A common focus of many studies is to examine why women are not represented in positions of power and authority. Our point of view inverts the perspective and studies the base of the pyramid, not the apex of the social division of work. The chapters in this book widen the social and political landscape within gender studies. Our concern is a feminized profession that is burdened by a lack of social status, which relegates women to work considered to be feminine.

This book does not aim to provide a detailed study of women teachers during a specific historical moment. Rather, our purpose is to integrate gender into the study of education. The chapters in this vol-ume explain the historical, political and cultural reasons for what Sandra Acker has called "the gender script," or the way in which the work of men and women teachers is influenced by societal expecta-tions about women's work and the caring capacity of female teachers when working with children.

There is a tension built into the study of teaching as a feminized profession. Many of the topics addressed in researching this phenom-enon involve teachers' work and equal pay, vertical segregation of the profession, difficulties women find in expanding their academic train-ing, low participation of women in teachers' unions, and in managing schools and school systems across different countries. Moreover, in an era of economic globalization, this tension increases as there is a significant growing drive of part-time employment among women in teaching (no benefits, no job security); rising differences in professional credentials between men and women in teaching; and segregation of women and men teachers by the age of the students they teach and the professional requirements for their jobs.

Our purpose in organizing this book is not to criticize or praise women and men for the choices they made to enter a feminized profession. Our main purpose is to further the knowledge of a pressing educational issue and highlight the existing inequality in the past and present organization of teaching as a "feminized profession." The chapters included in this book open a distinctive perspective on the historical and social research about the development of educational systems across nations. We hope this book will succeed in focusing attention on the gendered culture in the schools and the unique social and political events that led in each country to the universal presence of women in classrooms and to serious limitations in their autonomy and professionalization as teachers.

NOTES

1. Rosemary Deem (ed.), *Schooling for Women's Work* (London; Boston, MA: Routledge and Kegan Paul, 1980); see chapter 2 by Richards and Acker in this book for additional sources on history in Britain; see, Sandra Acker, "Gender and Teachers' Work," in Michael W. Apple, (ed.), *Review of Research in Education*, 21 (Washington DC: American Educational Research Association, 1995).
2. Rosemary Deem, *Women and Schooling* (London; Boston, MA: Routledge and Keagan Paul, 1978).
3. See chapter 2 by Richards and Acker in this book on history in Canada see also, Alice Prentice and M. Theobald, *Women who Taught* (Toronto: University of Toronto Press, 1991).
4. Josephine May, "Des—religieuses dans le siècle—et des homes de ce monde. Les élèves australiens de deux établissemts d' enseignement secondaire non mixtex se souviennent de leurs professeurs (1930–1950)," *Historie de l'Education* 98, Lyon, France: Institut National de Recherche Pédagogique, 2003: 167–185.
5. Nancy Hoffman, *Woman's "True" Profession: Voices from the History of Teaching* (Cambridge, MA: Harvard Education Press, 2003, 2nd ed.); Joel Perlmann and Robert A. Margo, *Women's Work? American Schoolteachers*, 1650–1920 (Chicago, IL: University of Chicago Press, 2001).
6. *Historie de l'Education* 98 (2003).

7. See Dan C. Lortie, *Schoolteacher: A Sociological Study* (Chicago and London: University of Chicago Press, 1975) and David B. Tyack and Elizabeth Hansot, *Learning Together: A History of Coeducation in American Public Schools* (New York: Russell Sage Foundation, 1992).

8. Perlmann and Margo, *Women's Work?*, 96–99.

9. David B. Tyack and Myra H. Strober, "Jobs and Gender: A History of the Structuring of Educational Employment by Sex," in Patricia A. Schmuck, W.W. Charters, Jr., and Richard O. Carlson (eds.), *Educational Policy and Management: Sex Differentials* (New York: Academic Press, 1981); David B. Tyack and Elizabeth Hansot, *op.cit.*

10. Perlmann and Margo, *Women's Work?* This same thesis is shared by Sari Knopp Biklen, *School Work: Gender and the Cultural Construction of Teaching* (New York: Teachers College, Columbia University, 1995) who shows that the teaching profession was constructed from a perspective of differentiated genders.

11. Ibid., 100.

12. Sari Knopp Biklen, *School Work* and Patricia Hill Collins, *Fighting Words: Black Women and the Search for Social Justice* (Minneapolis: University of Minnesota Press, 1998).

13. Jackie M. Blount, *Destined to Rule the Schools: Women and the Superintendency, 1873–1995* (Albany: State University of New York Press, 1998).

14. J. Camille Cammack and Donna Kalmbach Phillips, "Discourses and Subjectivities of the Gendered Teacher," *Gender and Education* 14 (2, 2002): 123–133.

15. See: UNESCO, Commission on the Status of Women, "Access of Women to the Teaching Profession," January 5, 1961, 9. The OECD (Organization for Economic Co-Operation and Development) publishes each year (in its Annual Report) the rates of feminization of teaching ("Education at Glance, OECD Indicators").

16. For the Dominican Republic see the chapter 8 by Juan Alfonseca in this book and for Brazil see chapter 5 by Fúlvia Rosemberg.

17. Steedman is referring to Froebel's dictum that "the ideal teacher of young children is like 'a mother made conscious.' " See Carolyne Steedman, " 'The Mother Made Conscious': The Historical Development of a Primary School Pedagogy," *History Workshop Journal*, 20 (1985): 149–163.

18. James C. Albisetti, "Catholics and Coeducation: Rhetoric and Reality in Europe before *Divini Illius Magistri*," *Paedagogica Historica*, 35, 3 (1999): 667–696.

19. Fúlvia Rosemberg, "Ambiguities in Compensatory Policies. A Case Study from Brazil," in Regina Cortina and Nelly P. Stromquist (eds.), *Distant Alliances. Promoting Education for Girls and Women in Latin America* (New York: RoutledgeFalmer, 2000), 261–293. Another policy of international agencies promotes the incorporation of young

adults with less than a high school education into the preschool level as well as primary and middle school levels. In Rosa María Torres and Emilio Tenti, *Políticas educativas y equidad en México. La experiencia de la educación comunitaria, la telesecundaria y los programas compensatorios* (México: Sep. 2000).

20. According to statistics from Tenti and Torres, *Políticas educativas*, in Mexico there are approximately 40,000 Community Instructors between the ages of 14 and 24 with only a high school degree or certificate.
21. See chapter by Sonsoles San Román in this book.
22. In those years, the Frente-Único pro Derechos de la Mujer was able to enlist more than 50,000 women; and many of these were teachers. On the participation of Mexican women teachers in feminist movements, see Anna Macías, *Against All Odds: The Feminist Movement in Mexico to 1940* (Westport, CT: Greenwood Press, 1982) and Regina Cortina (ed.), *Líderes y construcción de poder: Las maestras y el SNTE* (México, DF: Editorial Santillana, 2003).
23. Howard Gardner, *Multiple Intelligences: The Theory in Practice* (New York: Basic Books, 1993).
24. Michel Foucault, *Discipline and Punish. The Birth of the Prison* (New York: Vintage Books, 1979).
25. José Gimeno Sacristán, *La transición a la escuela secundaria* (Madrid: Morata, 1996), 111.
26. The American Association of University Women has published research documents on this subject, among which we should mention: "Girls in the Middle: Working to Succeed in School" (1996), "How Schools Shortchange Girls: The AAUW Report" (1992), and "¡Sí, Se Puede! Yes, We Can: Latinas in School (2000). More information can be found at the organizations's website: http://www.aauw.org
27. A recent study was done by Rosa María González, *Género y matemáticas: balanceando la educación* (México: Porrúa, Universidad Pedagógica Nacional, 2004).
28. The report "Gender Gaps: Where Schools Still Fail Our Children" points out: "In today's economy, women cluster in only 20 of the more than 400 job categories, and two out of the three minimum-wage earners are women" (p. 7). "Executive Summary" http://www.aauw.org/research/GGES.pdf
29. Lester Thurow, "Education and Economic Equality," *The Public Interest* (Summer 1972: 66–88).
30. A term employed by Mariano F. Enguita, "La tierra prometida. La contribución de la escuela a la igualdad de la mujer," *Revista de Educación*, No. 290 (1989), 21–41.
31. A term coined by Amitai Etzioni in the 1970s. See Amitai Etzioni (ed.), *The Semi-Professions and Their Organization: Teachers, Nurses, Social Workers* (New York: Free Press, 1969). Following the same line

of inquiry are Dan C. Lortie, *Schoolteacher. A Sociological Study* (Chicago: University of Chicago Press, 1975) and "The Balance of Control and Autonomy in Elementary School Teaching" in Amitai Etzioni (ed.), *The Semi-Professions and Their Organization: Teachers, Nurses, Social Workers* (New York: Free Press, 1969), 1–53.

32. Sandra Acker, *The Realities of Teachers' Work: Never a Dull Moment* (London: Cassell/Continuum, 1999).

PART I

The Daily Encounter of Teachers with a Feminized Profession

Professional Identities of Teachers during the Social Transformation toward Democracy in Spain

Sonsoles San Román

I have long been puzzled by a phenomenon I observe year after year as Professor of Sociology in the Department of Education and Teacher Training at the Universidad Autónoma in Madrid. There are so many female students in the classroom! Clearly, there had to be some explanation. My curiosity led me to investigate the historical and social origins of a phenomenon full of unknowns for me. I soon found myself immersed in dusty manuscripts and countless books until my efforts culminated in a study published in 1998.[1] Through my investigation of the origins and evolution of the woman teacher in Spain during the nineteenth century, I realized that there were social, historical, political, economic, moral, methodological, and religious factors that caused the professional expectations of Spanish women to be centered on teaching.

To clarify the meaning of "vocation," a term used by my female students to explain what motivated them to choose education, I embarked upon another research study. A second book appeared in 2001.[2] The focus groups that I utilized allowed me to understand the range of signifiers and social identities in the generations of women teachers who figure in the years leading up to the democratic transition in Spain.

My objective in this article is to examine the process of social change that led to democracy with the help of oral accounts from forgotten social agents: women teachers.[3] These women, who played a decisive role in shaping the personality, customs and preferences of Spaniards, are indispensable social agents in piecing together the story of one of the most significant events of the twentieth century: the arrival of democracy in Spain.

What were the cultural values of these women teachers? The answers to this question will help us understand the kinds of values that were inculcated in our childhood. In this chapter I will delve into the mentality of two generations of women teachers in order to help reveal the common characteristics of their professional identity. In this way I hope to show the steps, both forward and backwards, towards democracy through the analysis of meanings and social identities they assume with respect to their profession.

I am interested in showing how women teachers in Spain have assimilated and transmitted the social values of the transition. To this end, I will investigate the basis of their ideological formation; their self-images and consciousness in the advancing process of democratization; and the possibilities of transforming their social images, and the principles of their professional identity. Such a proposal requires a social and historical review that will allow us to separate the many strands of the present phenomenon of feminization, thereby situating readers in a more precise context.

This study, whose methodological basis is qualitative, should be understood with reference to a symbolic universe configured by a historical generation of teachers. The time period that I will survey is formed by two historical spaces that correspond to phases of historical developments in Spain in the second half of the twentieth century.

The first historical-generational space, which spans the conservative culture of the first Francoist government (1939–1959) and the immediate years after the end of the Spanish Civil War (from the 1940s to the 1950s), correspond to what is called the Generation of 56. The second historical–generational space in the 1960s saw the growth of a consumerist society and the appearance of pre-democratic, anti-Francoist attitudes. The distinctive experiences and attitudes surrounding democratization were lived by what is called the Generation of 68.[4]

This study focuses on two groups of women teachers: those who began to work between 1945 and 1955, during the first stage of the Francoist government immediately after the Civil War, and those belonging to the Generation of 68, who began to work at a time when Spain's pre-democratic consumerist society was being constructed in the 1960s.

TEACHERS UNDER NATIONAL CATHOLICISM

Historical Contextualization

On April 1, 1939, Franco announced the end of the Spanish Civil War.[5] During the early years of the Franco dictatorship the precarious political and economical situation in Spain turned back the pages of history. The social space assigned to women was once again reduced to the domestic sphere.[6] Female professionals suffered a profound setback. Under the pressure of Francoist ideology, women were concentrated in social spaces that were considered natural or appropriate for their sex: teaching, health, and service occupations. Women had a negligible participation in other areas that were better paid and more prestigious; these were reserved for the *head of the household*. Women's professional aspirations were sacrificed for the greater good of obtaining an unsalaried domestic worker. She greatly benefited society by assuming a motherhood that was not only biological, but social as well. Women, lacking both educational opportunities and the free time needed to study, were once again confined to the home. Their personal and professional development was blocked.[7] Sounds of protest were quickly heard: demands were increasingly made by women struggling for their rights in a century that the Spanish sociologist Victoria Camps has called *The Century of Women*.[8]

At the end of the 1950s Spain underwent important changes that would have an impact not only on its economic growth but also on society at large: the rural population abandoned the countryside in search of better opportunities in cities or in foreign countries.[9] Also, the silence of the dictatorship was broken with the protests of the Generation of 56,[10] which was the driving force behind a series of public demonstrations against the regimen that were quickly crushed. Several factors led to Spain's economic recovery. Foreign currency began to arrive: Spain had become the destination of choice for many tourists who, along with those who emigrated to other European countries in search of jobs, brought in needed currency. Spain's economic growth affected the lives of women, although at this point, only slightly.[11] Spanish society prepared itself for one of the most important changes of the century. The expanding industrial development fueled the consumerist society of the 60s and 70s, one that required female labor. The demand grew for women with a higher educational level, and important educational reforms were implemented in response to the economic, social, and political needs of the population.

In July 1974, Prince Juan Carlos was designated interim Chief of State, and on November 20, 1975 Franco died. Following the death of Franco the silence was broken, but fear would be slow to disappear. Spain, however, took its first steps on the road toward democracy. Women's associations and groups that had first appeared during the 1960s such as *Seminario de Estudios Sociológicos de la Mujer*, or *Movimiento Democrático de Mujeres*[12] tried to open the doors to exclusively male preserves, but from a clearly disadvantaged position. Spanish women struggled for their rights at a great distance from power and with the burden of a historical legacy that placed an almost insurmountable barrier in their way. Spain faced a problem with its mentality—a problem that could not be resolved by reforms alone. Undeterred, women began to speak of the constraints of domestic life and to voice criticisms.

Social, economic, and political changes required educational reforms to put an end to the obsolete apparatus of a society that had been afflicted by a dictatorship, repression, and the lack of freedom. On August 4, 1970 the *Ley General de Educación y Financiación de la Reforma Educativa* was passed, which once again allowed coeducation (which had been prohibited by the Franco regime after the Civil War), and presented an image of equal relations between men and women. One of the objectives of this law as to promote the education of women in order to respond to the needs of the marketplace. But the road was far from easy. Theory was one matter and putting it into practice was another.

IMAGES AND SOCIAL REPRESENTATIONS OF THE SPANISH TEACHER FROM 1950 TO 1980

Generation of Women Teachers under National Catholicism

Following the years of the fundamentalist National Catholicism of the 1940s, the Franco dictatorship emerged from isolation and autarchy, and gradually became integrated into the European culture of the era, a process that would take no less than 25 years. During this time, the issue of gender in basic education was once again addressed. On the one hand, Catholic fundamentalist ideology had shaped the mentality and sensibilities of the majority of women teachers who entered the ranks of teaching during the 1950s.[13] The special link of these teachers with early childhood education is closely intertwined with a complex idealization of the teaching vocation that is consistent with their background. But, on the other hand, the needs of an increasingly industrialized society demanded specific kinds of knowledge in the secular professions.[14] Both these demands and the attitudes of the women teachers which surfaced in the 1950s, represent a compromised training that would try to justify, once again, the unique suitability of women teachers because of their vocation, sensibility, and natural affinity.

This generation of women teachers were required to fulfill a neo-regenerationist[15] function by enlightening the rural population, but without distancing them from the regimen.

> I believe that during the past few decades, Spanish teachers have epitomized the traditional ethos of the petit bourgeoisie, and deepened it.[16]

In the 1950s the issue of women in teaching once again arose. There was an inherent contradiction, for while teaching itself was on the progressive end of the spectrum, women as a group were located on the regressive end. Women who entered teaching in the 1950s were progressive, to the degree that they carried out their duty to educate. However, the educative mission carried out by these women teachers was deeply inspired by the spirit of National Catholicism. They were charged with the regeneration of the people and were grounded in traditional values with fundamentalist overtones.

Ideological Poles

The discussion group—made up of five women teachers between the ages of 67 and 72, all of whom had started teaching in the 1950s—could be divided into two very dissimilar *habitus*,[17] as much for reasons of their distinct ideological orientations as their world of origin and points of reference:

1. In the traditional *habitus* that four of the teachers represented, the predominance of the affective pole over the intellectual pole was apparent. Traditional values of an affective type were intensely maternal. The discourse of the traditional *habitus* could be described as a predominantly feminine–intuitive discourse.
2. The modern *habitus*, represented by an Irish teacher who had lived many years in Spain, neither supported the feminization of the profession, nor the importance of vocation when entering the field. The attitudes and discourse of this teacher, who had a leftist political orientation and was active in the Comisiones Obreras labor union, was systematically opposed to the discourse of the traditional *habitus*.

Main Topics that Emerged during the Course of the Discussion

The teachers slowly began to take positions on questions relating to gender that were influenced by the dominant ideology of the 1950s. Of the five teachers in the group, four could draw on shared experiences. The other member was not only unfamiliar with the social context of the 1950s, but had also worked in a private school.[18] Although these biographical differences would soon produce differences in the group, there clearly existed a spirit of compromise that allowed them to reach a certain consensus. The teachers were at all times very attentive, and listened to each other with interest.

In the following sections, I will examine some of the social identities that emerged during the discussion held by this generation of women teachers.

Vocation

The social and political situation for teachers under National Catholicism shaped their vision, and as a result, their personal position about the teaching profession.[19] Their outlook was related to the gendered nature of teaching: it is a profession highly regarded for women, but lowly regarded for men. Such a vision reflected a society in which a woman followed a professional trajectory that was strongly conditioned by the family role she had been assigned.

At first the discussion was unfocused. The teachers began to take positions, however, after they were asked why they chose their profession. At the outset, they were concerned about how *socially correct* they seemed in front of their colleagues, and they defined a profession for which *natural feminine qualities* were required. In the process, they alluded to "vocation." The consensus was absolute among the women teachers with a traditional *habitus.* They affirmed in their answer that they had chosen the teaching career because "it was a pure calling. I loved it." "It was my vocation." Catholic ideology marked the lives of these women. They felt a personal calling to save their country in their role as legitimate representatives of the culture of the regime. It is important to remember that the women and men teachers who were trained under the Plan of 1936 of the Second Republic had been purged from teaching because they would not offer classes on religion and did not profess Catholicism.[20]

The idea of "vocation" was presented to an entire generation of women Catholic teachers as a kind of predestined path requiring a spirit of sacrifice toward others. The participants in the discussion group defended the high rate of women in teaching. They believed a woman possesses the natural and ideal talents for becoming a substitute mother,[21] pointing to her natural maternal instinct, her tenderness and devotion to children, and her ability to communicate in a way a rural population would understand. They cited a woman's superior capacity to adapt, as shown by their own adjustment to the abject conditions in the poor towns where they worked in post–Civil War Spain. A woman's innate spirit of sacrifice and generosity was also mentioned. Despite the lack of job prestige, the low salaries they received for years, the moral strictures placed upon them, and the lack of freedom that they endured, these women teachers did not feel exploited, nor were they conscious of the professional limits that were imposed on them because they were women.

In their discussion they limited themselves to telling anecdotes about their experiences. Significantly, they were not able to link their professional life with the political situation in the Spain of National Catholicism, nor did they understand the deep impact that political reforms and social change had on their teaching practice. In short, they did not show any awareness of the process of transition. In framing their job within the parameters of personal responsibility and selfless commitment, they did not feel exploited at work. Their sense of "vocation" led them to face obstacles with a spirit of self-sacrifice and duty, and to put up with dreadful working conditions. These were women who had been educated within a society in which reflection and criticism were considered the enemies of religious dogma, and they accepted the limits placed on women by the Francoist regime.

When asked which professions were suited for women, the teachers with a traditional orientation mentioned teaching before all others, followed by nursing, pharmacy, and the complete range of the humanities. These teachers identified themselves with the completely feminized image of teaching in Spain, which was determined by the subordination of the woman teacher to her husband in the development of their respective careers.

The progressive teacher, who had taught in private schools and was unfamiliar with the rural Spain of the 1950s (unlike other members of the group, who had worked there), showed more of a critical distance. Her discourse tended to be more analytical and reflexive and she denounced job exploitation. Here was an activist from the *Comisiones Obreras* labor union who was perfectly aware of the ideological framework that shaped the course of education starting in the second half of the twentieth century. Having experienced social changes on the level of the classroom and the street, she was aware of the cozy relationships established between politics, education, and religion during the Franco regime. The contrasting viewpoints she interjected enriched the debate.

> Another aspect of the profession that affected me greatly was the miserable, wretched conditions and the fact we had to work at a very low salary. This politicized me as to what was happening in Spain. I became interested in social problems and union problems, which I personally experienced. I've always been active in the union. I was there from the beginning of what the union is today.[22] That's how I got to know many of my teaching colleagues, when we tried to start the union; it was an important experience for me.

In a moment of heated debate, facing a unified group of Catholic women teachers, and conscious of the influence of social factors on their choice to become teachers, her assessment was blunt: "Destined? No way!" Instead of a biological condition, she saw the seemingly natural association between womanhood and teaching as a social destiny. The idea of a vocation was exposed as a trap, a form of labor exploitation, an artifice intended to promote conformity and submission in women teachers of that era.

> I also object somewhat to the word vocation, even though of course I love my profession. I believe it's a rope that's used to hang teachers, especially women teachers: you can't protest, they don't have to pay you well, you can't do anything.

The traditionalist bloc didn't share this belief; in their view, it undermined their professional image. "Vocation" is the term they used in presenting themselves as good professionals, or what is nearly the same, as women suffused with the religious atmosphere of National Catholicism. This association led them to argue that a woman's female qualities made her especially suited for teaching. They looked for new arguments to link the word "vocation" to an exemplary performance on the job, thus rejecting the negative image presented by the Irish teacher.

When the dialogue was well underway, there was a moment in which terms were reconfigured and new meanings were sought. A more relaxed mood prevailed. The participants' initial uneasiness dissipated as a sense of trust grew among the teachers. They were no longer intent on proving that they had done good work in their chosen vocation; rather, they wanted to hear the experiences of the other group members. Eventually they came round to considering the social factors that led them to choose education: a lack of opportunities to pursue studies in other field due to financial constraints, or the lack of universities in their rural areas. And as the discussion continued, they did not hesitate to openly affirm, without resentment, that religion was an obligation imposed on them.

Social Change

For the traditionalist teachers, the veil that concealed the mechanisms governing women's entry into teaching also blocked their awareness of the changes around them. Although at first they stated, "it happened without our realizing it," they later allowed that changes in their teaching style may have occurred due to the importance they placed on in-service courses and training: "We've taken many courses in teaching mathematics, language, in a different way. . . ."

Once again it was the teacher with the progressive outlook who pointed out the importance of social change during the transition toward democracy. This transition was accompanied by a significant transformation in teaching methodology. Learning through memorization was discarded in favor of reflective and critical learning strategies, in keeping with the wider access to culture and education. These transformations within education were experienced and conceptualized in a different way by the two ideological blocs. The traditionalists, who professed no awareness of a changing society, merely indicated that developments in education in their over forty years of teaching were a result of in-service training, the loss of the teacher's authority, and the different attitude of new teachers. While the progressive position had a high regard for the reflective autonomy of the students, the traditionalist view was more ambivalent:

> Well, one very big difference is that you're no longer the one who always knows the truth. Instead, you're questioned on everything. For example, you can't say "because the teacher says so." If they go home and say the teacher wants me to copy this word twenty-five times because I misspelled it, the mother can say, "Twenty-five times! No way! Five times is enough." This has happened to me.
> . . . with the arrival of democracy, parents and teachers, everyone has been evolving, changing. I don't know if it's good or bad, but it's different.

The progressive attitude is quite different:

> I'd say that one of the biggest changes is that the teacher has stopped being the authority, the one who knows everything. . . . A colleague

told me something amusing the other day. He explained something in class and they discussed and commented on it, and then the next day a boy came up to him, a kid in the seventh grade, "what you said yesterday in class is true because I heard a man on television say it."

The group continued its debate. The moderator asked whether there had been a moment when teaching methods changed. Surprisingly, the Francoist Catholic teachers still continued to show no awareness of the social change that had taken place. Once again, it was the progressive teacher who pointed out the importance of social change in the transition toward democracy and in the General Education Law of 1970.

Feminization

The group tried to describe the possible causes that produced the strong feminization of the profession. As this topic was discussed, the two contrasting positions resurfaced. While the significance of economic factors was argued by the progressive teacher, the traditional teachers emphasized, at first, the specific nature of gender relations in classroom exchanges and communication. By implication, they affirmed gender stratification: in the subconscious of these Catholic teachers who had been educated in Francoist ideology, men, as such, possess a superior status. At the same time his status gives his work a higher value on the labor market, it also distances him from the lower levels of the educational pyramid.

While women teachers with a traditional *habitus* (who stated that their choice of professions was motivated by vocation and a love for children) strongly defended the feminization of the profession, the progressive teacher once again defended a professional vision in which the educational process was independent of the question of gender.

The traditional bloc understood that teaching implied no rupture with a woman's customary role. However, it was not a man's natural place. "It was said that if a man wasn't good enough to do anything else, he stayed in teaching." They supported the argument that the innate qualities of women made them better teachers. "Men of my generation remember male teachers as somewhat aggressive. They tended to be more disagreeable. A woman teacher, in comparison, was wonderful."

The progressive view, however, reasoned that if the genders had different natures, it was due to purely social, not biological, reasons. "It's because the times used to be more authoritarian. Men had a role to play and women were not regarded as highly in that role. It's important to have both sexes working in the profession." In this way, the professionalization of teaching seems to be linked to overcoming discrimination based on gender.

The traditionalist teachers disagreed with this view. They asserted that the method employed by men teachers is more authoritarian, and for that reason, less compatible with a child's nature. "I saw that their characters were hardly feminine." "We had two in my school. They were assigned to young children and they didn't last." Although they believed they were praising feminine nature, they were reinforcing the dominant masculine and discriminatory ideology: "Men don't have the right character for children."

These affirmations of female attitudes is reinforced by attributing fundamentally passive and adaptive values to women: "It's a question of having a little bit of patience, a little bit of order. You have to keep all your work in order." Or that "women are more skilled." The progressive teacher, on the other hand, didn't believe that "men teachers were more inflexible in their way of thinking."

The progressive *habitus* adopted a democratic outlook and offered a vision of the educative process that was strictly professional:

> [I]t's good that both sexes are in the profession . . . it's often said that "the person who can't get into a university goes to Normal School," which is absolutely incorrect because it depends on where you live and your financial means. I have always defended admission standards as a positive thing for teaching, because in rural families the boy who was studious would either go to the seminary or to a Normal School.

Professional Image

The teachers from the traditionalist bloc told anecdotes about their experiences in the Spain of the 1950s. Their stories about their first assignments in rural villages reflected the ambiguity between their professional image and their status as women:[23]

> One day the servant of a big landowner asked me if I would give classes to the young men in town. I thought, under no circumstances! To go to their house at night! . . . Then an official of the town appeared with a group of the young men, to request classes. They promised they wouldn't touch me, and that they would pick me up and take me home. And it was really one of the most beautiful experiences I've ever had in my life. To be with young men who were so rough, so unpolished. I taught them how to read . . . and, of course, it was an act of charity.

The teachers with a conservative outlook expressed approval of those times in which "the teacher had some say over religious matters. Back then you went to Mass with the students, you played an important role. . . . you did many things with the priest." However, they didn't fail to mention that religious practices "were obligatory." The teachers, who all shared the same religious beliefs, were aware that their function was subordinate to ecclesiastical authority.

As teachers, they were professionals; but as women they were dependents. They were pressed into service as the need arose. Town authorities knew how to take advantage of any contribution these women could make:

> "We had to act as priests, even."
> In my first school I was the veterinarian, the doctor, I gave injections. . . . I bought medicine, I even accompanied family members who had lost a loved one. . . . in the first town I worked in, I arrived and there was a wake for a child. . . . I met the whole town there. . . . this has stayed with me my whole life. Wherever you went someone had died, and there was Nieves praying the rosary. . . . I never had problems with anyone.

The woman teacher represents a subordinated authority. She is a focus of criticism and scrutiny derived from two circumstances: as a professional, her duty is to educate the rural population, but as a woman she is an object of desire and a role model. In the rural enclosure, she is the center of the villagers' gaze.

> [W]hen I got off the horse someone always helped me. If it was a man he always grabbed me by the waist. I've always been very ticklish. . . . One day Don José said to me, "Miss, you're the only show, the only theater, the only entertainment we have." You got on and off the horse. . . . Everyone watched what you did. . . . you had to be very careful about what you did, what you said. . . . Everyone looked at you. You were a mirror.

The woman teacher, the town's mother, turned into an object of desire. Her desirability became more intense as she became more unattainable:

> I don't know if I should tell this story: in another town where I worked . . . I was living in the house of the other teacher. . . . she was a widow and she asked me to stay with her, to keep her company. . . . One day another teacher, a man, told me, "Be careful to close your

window curtains when you go to sleep. There's a group of young boys who stand by your window to watch you undress." Another time, a man who knew my father threatened that he would tell him that my boyfriend was writing to me every day. The postman brought me his letters. This man thought it was wrong and he told me, "Look, if your boyfriend keeps writing to you every day, I'm going to tell your father."

WOMEN TEACHERS OF THE GENERATION OF 68

The duties assigned to women teachers under National Catholicism were restricted to instilling religious values and the necessary attitudes for the consolidation of the political regime. The women of the Generation of 68 were protagonists at the outset of modernization in Spain. When Spain came out of isolation, new expectations for education were developed in an attempt to guarantee a qualified supply of labor.[24]

At the end of the 1960s educational policies underwent significant reforms to meet the demands of modernization. Spain was opening itself to the outside world. Politically speaking, the situation required not only that ministers be replaced but that the design of the educational system be overhauled so that a citizenry would be created that could confront the social challenges of modernity.

In July 1962 the first plans for the educational development in Spain appeared.[25] The country needed to lift its cultural level to that of neighboring countries. Levels of achievement in the schools had to be increased to combat poor reading and writing skills and produce a qualified labor pool. More schools were necessary to serve school age children whose families had migrated from rural areas. Repercussions were felt in the professional requirements to become teachers and in the course of study teachers had to follow during the years leading up to the democratic transition.

Women's groups and feminist movements returned again to the country in the 1960s. Spanish women started to take steps on the path toward recovering lost rights and gaining new ones. They fought for the right to participate as citizens in both public and professional life on an equal basis, a status denied them because of what were seen as incontestable and divinely ordained differences.

Women teachers of the Generation of 68 lived through this era in their double condition as women and as educational professionals. Responsible for following the new curriculum set by the government, they felt close to the strength of tradition in Spanish society. During

Spain's transition toward democracy, however, they experienced political change and adapted themselves to the new political ideology.

The Spanish women who were heirs to May '68 in France asked that women be given independence and freedom, both professional and sexual. From their discourse it was clear that they were not willing to bend to external pressures. However, they acknowledged that on more than one occasion they concealed themselves behind their female mask, hiding their way of thinking, to avoid conflicts with the authorities, students' parents, or with rural populations. These women teachers participated in strikes and protest marches for higher salaries, were involved in the educational reform movements in the 1980s, and were imbued with the spirit of Rosa Sensat's pedagogical reform.[26] They were not conformists; they were combatants. Without the religiously inspired spirit of sacrifice of their predecessors, they dedicated their energies to changing society through the school. In their opinion, their efforts have not been rewarded. They have suffered through strikes, persecutions, and the repression of a dictatorial regime, but they have also seen how their lungs suddenly breathed in the fresh air of freedom. They resisted the control of authorities such as principals and school inspectors, and bravely introduced new practices in small rural towns limited by the customs of their inhabitants. These young innovators were, in the most literal sense of the phrase, the first "to wear the pants," without knowing that by doing so they were utilizing a symbol that stood for equality between the sexes.

During their discussion, a society in the throes of change appeared: tradition and modernity; a before and after; the problems presented by the massive increase in school-age children on the outskirts of the big cities, the result of the country's internal migrations; their subordinate role in the family and in society; the rigorous control they were subjected to in small rural towns. Without a doubt, they were the main protagonists, and occasionally the victims, of a series of educational reforms introduced by the State during the difficult time of transition toward modernity. From a historical and social perspective, their accounts are an obligatory point of reference in order to understand this episode in the history of Spain.

They reflected deeply while attempting to locate through group consensus what their identity was as women teachers; they looked back on specific situations in which their social images were transformed as they exercised their profession, and searched for explanations. Their attitude was very open and a dialogue was fostered. They listened attentively, asked each other questions, and gave each other the floor.

Ideological Poles, Attitudes, and Social Images of Women Teachers in the Decade of the 1960s

The discussion group, made up of five women teachers whose ages ranged from 49 to 58 years old. Some had attended Normal School when its curriculum was set by the Plan of Study of 1950; the younger teachers had been educated under the Plan of 1967. As a group they displayed a high degree of mutual support and respect. Within this group of teachers from the Generation of 68, two distinct *habitus* could be discerned, represented by two models of womanhood. These models were rooted in the contradictions inherent in a pre-democratic Spain that had opened itself to modernity and set in motion the transition toward democracy.

A managerial *habitus* corresponded to three women teachers who had been educated under both the 1950 and 1967 Plans of Study. They held administrative positions in their schools or were affiliated with the Comisiones Obreras labor union. Their introspective discourse was consistent with the model of a non-conformist woman who does not hesitate to denounce the subordinate position she has held as a female professional. These teachers even questioned the extent of the democracy. Their politicized profile allowed them to visualize everyday problems in the classroom from a certain distance and to reflect more deeply on the relationship between the classroom and society at large. Reflective and critical observations were woven into their discourse in addition to narrative and anecdotal accounts.

The emotive *habitus* was represented by two women teachers who displayed a more limited and emotional vision, one that was focused on day-to-day problems in school. Unlike the teachers of the managerial *habitus*, they had never left the classroom to take on administrative positions. They showed a higher degree of conformity when difficulties arose with those in authority or the inhabitants of rural villages. Their language seemed tailored to the classroom and they displayed a submissive, timid attitude within the group. In the emotive *habitus* two kinds of discourse can be observed simultaneously: the anecdotal and the narrative on the one hand; observant, enthusiastic, and empathetic, on the other. These differences divide the emotive *habitus* into two factions: the modern faction, represented by a woman teacher who we could say was from the "generation of the disillusioned"—pessimistic, and longing for the days when the teacher was an important member of the community. She reinterpreted and reformulated, respectfully and with interest, the ideas put forth by her colleagues with respect to the social changes that occurred en route to democracy. The traditional faction was characterized by a factious, defensive, and more domineering teacher, who showed a greater degree of assimilation to the patterns of behavior, which were required of women in the ideological context of a traditional society.

With this generation, a renewed focus on the professionalization of teaching, lost with the outbreak of the Civil War and the abandonment of Republican ideals, is seen. This trend, however, did not reverse the feminization of the profession. Teaching continued to show very high levels of female participation at the lowest levels of the educational system.

Topics Which Emerged during the Discussion

Unlike the first group of teachers shaped by National Catholicism, who in the initial stages of their discussion sought to reach a consensus, no similar agreement was achieved between the teachers belonging to the Generation of 68. The "vocational overtones" of the profession—which were affirmed by the teachers in the previous generation—were not professed by this younger group, who from the outset began to position themselves clearly within the dominant ideology of pre-transition Spain. They emphasized the lack of opportunities for women from small rural towns to pursue studies in any other field but teaching. In most cases, they had been selected by the women teachers in their schools, had obtained a scholarship, and could count on the blessings of their family who, having sacrificed their savings, expected their daughters to enter a career that would guarantee their investment. Given these restrictions, teaching was their first and only choice.

Although none of the teachers in this generation used the term "vocation" in describing the motives behind their choice of profession, the majority of the teachers of the managerial *habitus* referred to their early vocation, "*since way back*," when locating the origin of their desire to become a teacher. Expressions such as "*it was very clear to me*," or "*I took it for granted from an early age*," were used by teachers who had decided on their vocation since childhood through both family and social role models.

When we compared the managerial *habitus* with the emotive *habitus*, we discovered, unexpectedly, that the women teachers who showed the least interest in becoming teachers, and were even at times resistant, were the ones with the greatest sense of vocation in the group:

> It never entered my mind to pursue a career in teaching . . . it coincided with a financial problem. . . . I began to work and from then on things changed. . . . I was very young, eighteen years old, with so much freedom, so many opportunities, it was fantastic. . . . and suddenly I realized that I liked teaching. I've never thought about leaving it.

The teachers of the Generation of 68 didn't wrap themselves in the word "vocation" when identifying themselves with a profession that the previous generation of teachers had seen as "socially correct." However, they were still subject to work environments based on gender. The political, social, educational, and economic changes brought on by the transition toward democracy were not able to overcome the legacy of tradition. In the reflections of these teachers, teaching was still regarded as a female career.

By distancing themselves from any religious overtones, these teachers cast off the sense of duty and self-sacrifice that had clung to this so-called sacred work (in which performing one's job was seen as a religious obligation).[27] More importantly, they began to insist on standards of professionalism. Nevertheless, the concept of vocation was unexpectedly introduced at one point by a teacher of the emotive *habitus*, who referred to the notion of a "romantic relationship." With this phrase, the teacher signaled a personal commitment; a challenge that would be greatly rewarded; a sentimental relationship between the woman teacher and the child. Education was now envisioned as a way to promote and advance social mobility, and the teachers knew that their function was to lead a literacy campaign that would let Spain take its place with other industrialized nations and leave stagnation behind. The doors to modernity were beginning to open.

> Romantic for me means vocational. . . . someone who says "I fight for the nation, I fight for the people." . . . I was nineteen years old. . . . And yes, I had romantic ideas, I threw myself into it. That's what "vocation" means. Look, I had rejected it before. It's a job you need a special calling for, absolutely, because you get totally and absolutely involved in it as a person, not just as a professional, but as a person.

The teachers belonging to the managerial *habitus*, who were somewhat removed from the emotive discourse, refer again to vocation as they reflected back on difficult moments in small rural towns, where their early vocation was reinforced by the satisfaction of doing good for the community.

> We read all of Rosa Sensat's books. . . . We were always fighting for something. One time we were robbed. We had formed a parents' association. They donated what they could, and with that money we bought a movie projector, a super 8. We did a few interesting things, we had a TV. . . . after the robbery was discovered, the parents of our students came. . . . A half an hour later everything was back. "You don't need to tell the cops. . . . No one can break into this school, dammit." It was as if they had been robbed. . . . I'll never forget experiences like that, at school, with "my kids."

The Ascending Line of Professionalization

The consensus of the group returned when they affirmed that a sense of vocation was not enough: "one must be a professional." Most of the teachers did not have a high level of education, and their teaching practices were deficient. They had received their training as teachers under the Plan of Study of 1950, which in many cases lasted only a week. The interviews yielded ample information about the archetype of the woman teacher from this generation. With their brand-new teaching degrees they left their towns for small rural villages. There they were mistaken for students, even by the school inspector himself. They were too young and inexperienced to command respect among the farmers who closely monitored them: "They never treated me as usted."[28]

Young, insecure, and constantly under the scrutiny of the rural farmers and the authorities, they were used to a way of life that clashed with that of the village. They faced the challenges of their first teaching assignments with little preparation. In contrast, the more extensive professional training provided to new teachers today met with their approval. Their own lack of training was, in their opinion, a great wrong. These early experiences that were deeply engraved in their memories, gave them a down-to-earth understanding of the realities of education in Spain. They accepted this reality and distanced themselves from the term "vocation." With both feet on the ground, they demanded better academic preparation to become educational professionals. The lack of training was evident in their personal accounts. They confessed to feelings of shame when they realized that, beneath a superficial appearance of security, there were tremendous gaps in their academic and professional knowledge. These teachers had tough characters and a thick skin, and had learned to resolve difficulties as they went along. How could the idealized concept of "vocation" have any meaning in their discourse? Training and professionalization was all they asked for.

> For many years I didn't want to tell this because it embarrassed me, but it's a sign that I've put it behind me. . . . I drew a circle on the blackboard once and I told them, "Children, this is an 'O.' " . . . the training I was given was of little use.

In most cases these 18-year-old girls simply imitated the maternal habits of the women teachers in their schools. These teachers did not distinguish between private and public space, and were not interested in their students' advancement:

> I arrived there, and I saw that the teachers brought their knitting to school, so I bought some knitting needles. They told me, "Bah! The stupid ones are stupid, and what can you do about it?" They gave the students some simple math problems and then they began knitting. I arrived there . . . without knowing what to do in front of the classroom. . . . and I copied what I saw.

These kinds of experiences were common among the teachers who had close contact with the miserable living conditions in Spanish towns in the 1960s. With the changes that occurred in the educational panorama during the 1970s, these teachers' social conscience grew, causing them to break definitively with the ideology of the previous generation of teachers under National Catholicism. The theme of modernity established itself in the discourse of this generation, which experienced at first-hand the effects of the internal migration from rural to industrialized zones, the problem of overcrowded classrooms, and the appearance of new demands once the democratic transition was initiated. These teachers, for whom nothing came easily, worked in marginal neighborhoods and impoverished rural zones, and valued a right that had been denied them: professionalism.

Feminization

The topic of the feminization of teaching was introduced in the group by one of the teachers of the managerial *habitus*. Curiously, none of those convened wondered why the groups were all-female. Force of habit made them assume it was natural for there to be no men present—they weren't present in classrooms, either. The small group discussion simply reproduced their day-to-day life at school. In this feminized turf, the woman teacher may feel comfortable, but she also at times feels stifled. This discontent led one of the teachers of the managerial *habitus* to abandon the classroom:

> It was suffocating, I was tired of it. I worked in a pre-school with four other women, who had worked together I don't know how many years, and the young children. They ran out to the patio: "Miss, Miss, Miss!" And the other teachers couldn't talk about anything except school, or that their child had the flu, or that they forgot to turn the washer off or that it's cloudy today. The moment came when I couldn't take it any more, and I made a complete change.

The rift widened between the two orientations in the discussion. The teachers of the emotive *habitus*, who did not leave the classroom for administrative positions, did not regard feminization as a problem. They either did not comment or feigned agreement in discussing feminization. On the other hand, the teachers of the managerial *habitus*, who were more ambitious and conscious of the limitations implied by being woman and working in a feminized, devalued space, brought up the negative effects this phenomenon had on students. They treated feminization as a problem that needed to be resolved.

> It's a problem. The male figure is important for children, just as important as the female figure. For me it's fundamental. I think that school has to be a reflection of what life is like, what society is like, and children have to see male and female behavior. And they have to see that a male teacher leans over to tie their shoelaces just as a female teacher does. They shouldn't see teaching as a women's ghetto. Besides, I think that this exchange is good for women as professionals. An all-male group, or an all-female group, is not as rich as a group that has males and females who share experiences. . . . Children have a different attitude, because the relationship with their teachers is different. A young girl can come up to me and hold me like this. But probably with a man she won't do that at first; there's a different sensation. So that teacher will try to create the same feeling in her, but the relationship between them is different: the tone of voice, the posture, everything. I think that children have to see this, experience this. They should be part of it.

The topic arose when as moderator I asked, "Do you work with only women in elementary schools?" A teacher from the managerial *habitus* instantly expanded on the subject, joined by the rest who either agreed that yes, all the teachers were female, or that they saw the classroom as a space for women.

The interview not only elicited the opinion of teachers from the two *habitus* concerning the feminization of teaching, but also revealed the personal reasons that prompted the teacher who lamented the lack of men. From childhood on she was accustomed to associating men with teaching, as her father had been a teacher. In the interview, the teachers with an emotive *habitus* discussed the qualities of women for education, pointing out that teaching was a promised land for women:

> I'm embarrassed to say it, but I've always thought that women teachers are better. . . . When a child has a male teacher, you think, that's great! because there are so few of them. . . . but expectations are higher at the beginning than at the end of the semester.

Others were less blunt:

> I think there are so many women in teaching because the role of the mother and the role of the teacher was confused. Teaching is something that's more maternal. . . . The working hours coincide with your children's schedule.

The teachers with an emotive *habitus* limited themselves to describing the situation they personally encountered: "The classes are coed, and yes there are many more women. Men are leaving." However, feminization had been raised as an issue and then questioned; signs of modernity continued to appear in the discussion. This generation discarded the ideology of the teachers of National Catholicism, who defended, in order to reinforce their identity, the feminization of teaching as a part of the natural order.

Social Control

This generation of teachers, who experienced religious repression, sought ways to avoid teaching religion, a required subject in the school curriculum. Their objective was to separate school from the determining influence of the Church, which they had experienced at close range. Indeed, the teaching of religion was not taken on as a duty by the Generation of 68:

> We had to teach religion just as we had to teach penmanship. . . . it was one more class. Students had to pass it and that was it.
>
> Back then everyone had to be Catholic in Spain.
>
> I had to teach religion against my will.
>
> I'm not a believer at all.

When they began to work they were put under strong control. All of them had to swear to the Falange[29] after passing the teacher's examination: "If you didn't, they wouldn't give you your degree. . . . the janitor told me, 'Shut up and take the oath!' " They also had to take a course in physical education for girls: "You had to wear a special striped skirt . . . made with a special fabric." They were required to work for a time in social services. These measures were in part to guarantee their femininity, if not their professionalism and their loyalty to the regimen. In the villages where they were assigned, their conduct continued to be closely monitored:

> They thought you were a whore, more or less. There was no television. I think now everyone has a television and has seen people in all different ways, including naked, but back then nobody had a television. . . . Mothers would come into my classes and they stood outside the school . . . they didn't like the way I dressed and the way I was. They complained about me. . . . The inspector discovered the real reason, and said, "I think it's because they were used to a much older teacher." And so I called a meeting. . . . and from then on, things began to change. They began to see me in a different way.

For these teachers, wearing pants—a symbol of masculinity— became a symbol of rebellion against the social control that required their spotless behavior. With no law to protect them, the women offered each other help and appeared in groups wearing pants so that no one would be singled out for criticism when the repressive educational authorities or school inspectors insisted that they were not wearing proper attire. It was a challenge to authority, against a social model that was beginning to change in Spanish society. Their rebellious conduct reflected a society that now had the necessary strength to confront the social change that would lead toward the democratic transition.

> My first Inspector was a man who was already retired. Because I was wearing pants, he immediately said to the principal that she should give me only two pieces of chalk.
>
> [T]he Inspector didn't like women who wore pants. Back then I was young and outspoken and said, "When you show me the law against it, I'll change." But in the end, I didn't wear pants that year.
>
> The first time the Inspector had a meeting with us, it was to tell us that we shouldn't wear pants, that he liked us in skirts. He added that we shouldn't go to nightclubs, and that if a man looked at us, we should lower our heads. For me, my first experience with this inspector was, just imagine, how can I put it . . . a little testy.

I appeared in a mini-skirt, with braids and a cigarette in my hand, very bold. Everyone in the village was shocked. There was no television, no radio, there was simply nothing. They were totally isolated there. And so they started to keep an eye on me, to monitor me, to the point where they even came into the school and stood there. And I said, "I'm about to start." And they said, "We're not leaving." They treated me like a child; they never treated me as "usted," nothing like that. And so they stayed there a few days, watching what I did. And finally they lodged a complaint against me. Yes, a formal complaint . . . they didn't understand me. I arrived there and I didn't fit in at all, I was someone they couldn't see as a teacher.

QUESTIONING THE VALUE OF DEMOCRACY: EXCLUSION AS A PRINCIPLE OF IDENTITY FOR WOMEN TEACHERS

This generation, which was immersed in a whirlwind of legislative changes affecting education, openly criticized reforms that were "imposed" without offering in-service training to teachers. "We're the only part of the public or private sector that doesn't retrain its workers when laws are changed." Their consensus on this point strongly united the two *habitus*, and their feelings of exclusion emerged as a new form of identity:

[T]hey have no idea about the reality of schools. They make a law full of spirit and rhetorical words and they send it to you. . . . "now you have to learn it" [ ] The same thing has happened with all the laws we've had. . . . absolutely all . . . your experience isn't important . . . your opinion about the law doesn't matter.

The teachers felt like marionettes whose strings were pulled by the State in order to achieve the cultural model it sought through its educational workers. These teachers did not feel valued as professionals, and their criticism was sharp. Their political conscience became manifest as they denounced those in power for not seeking their opinion. They were aware that the cultural models and curriculums they teach in school were imposed. Because of these grievances, they began to question the principles of democracy, and speculated on the different strategies employed by those in power to avoid teachers' input in formulating public policy.

The reforms introduced by the 1970 *Ley General de Educación* have also worn away at their professionalism. Its requirement to prepare individualized reports on each student overwhelmed teachers, causing them to go home after school with 40 to 50 students' files that they had to submit the following day: "the 1970 Law . . . reports, reports, reports. . . . you gave the students back their reports the next day, and then you realized . . . that the children hadn't even looked at it. . . . you wasted your time." The feeling of helplessness was absolute. "We operated on a trial and error basis. If you did something wrong you told the others so they wouldn't make the same mistake."

The Law placed special emphasis on written corrections, which implied a more indirect contact with the student: there were fewer oral and group activities. The idea of group work in a class would be taken up 20 years later with the LOGSE,[30] which again imposed unreasonably high demands on the teaching staff.

Reflections

The connection between the sense of duty, vocation, and religion that Weber established is clearly relevant in understanding the contrasting mentalities of these two generations. In *The Protestant Ethic and the Spirit of Capitalism*,[31] Weber demonstrates the degree to which religious beliefs govern a person's behavior, and more specifically, how they determine the sense of duty that make an individual perform his or her profession with dedication and devotion. In this sense, the importance of education is underscored, as it shapes the type of occupations that a social system requires to function economically.

If we apply Weber's investigations into the origins of professional duty, it is possible to unravel the meaning of the term "vocation." Indeed, by exploring what "professional duty" means to two generations of women teachers, we discover the degree to which religion has been utilized to inject irrational elements that shaped their professional choices. Thus, the generation of teachers formed under National Catholicism used "vocation" as a means to establish their credentials as good professionals who had fulfilled their duty, made many sacrifices, and devoted themselves to their profession without seeking financial gain. They described their vocation, which was tinged by the religious overtones of National Catholicism, as one that required dedication, patience, and a sweet nature. In keeping with the educational model that relieved women of political responsibilities, they avoided speaking of politics, a topic that was considered socially incorrect by this generation of teachers. In contrast, the teachers of the transition generation showed a clear ideological shift in their discussion. They distanced themselves from the Church, which had lost the importance and ascendancy it traditionally held in Spain, as shown by the significant decrease in the number of religious vocations. Thus, to the degree that religion has lost its importance in Spanish society, the term "vocation" has disappeared from the discourse of the teachers of the Generation of 68. They did not feel they were "called" to be teachers, nor were they submissive. They did not assume the dogma of the Catholic faith. Having abandoned the spirit of priest-like sacrifice and obedience that characterized the previous generation of teachers, this generation insisted on their professional rights and demanded better salaries.

NOTES

1. Sonsoles San Román, *Las primeras maestras. Los orígenes del proceso de feminización docente en España* (Barcelona: Ariel, 1998).
2. San Román, Sonsoles, Ed., "La maestra en el proceso de cambio social de transición democrática: Espacios histórico generacionales," Informe publicado por el Instituto de la Mujer (Ministerio de Asuntos Sociales), Madrid, Colección Estudios, 2001.
3. For a more complete account of how the focus groups of women teachers were formed and how their oral histories were elicited, see San Román, "La maestra en el proceso de cambio social," pp. 19–31.
4. A third space, which corresponds to the culture of democracy following the consolidation of the constitutional state and the normalization of the parliamentary government and its values (which is articulated in the Generation of 1981–1982 after the failed coup d'etat on February 23, 1981, has been examined in San Román, "La maestra en el proceso de cambio social," pp. 291–320.
5. As is common after a major conflict, the Civil War had dire economic consequences for Spain. On June 5, 1939 Franco announced that Spain would be reconstructed under an autarchical regime.
6. Before the Second Republic of 1931, Spanish law was characterized by discrimination against women in the realm of politics, work, and family. The reforms made between 1931 and 1936 favored the integration of women into the labor market in professions that went beyond those considered by society to be feminine. Sadly, the Franco regime put an end to this initiative and once again reduced women's professional options.
7. This historical legacy is still present in the mentality of the Spanish people. Much needs to be done to achieve equality between the sexes.
8. Victoria Camps, *El siglo de las mujeres* (Barcelona: Cátedra, 1998).
9. Foreign migration was very intense throughout the twentieth century. Until 1961 the destination selected by the majority of emigrants was the Americas. Between 1963 and 1973 the yearly average of emigrants to Europe was higher, but by the early 70s the Americas were once again chosen by a majority of Spaniards. Ninety-eight percent of all Spanish emigrants went to either Europe or the Americas. See Abdón Mateos and Alvaro Soto, "El final del franquismo, 1959–1975: La transformación de la sociedad española," *Historia* 16 (1977): 22–42.

10. These demonstrations exposed the contradictions in the country's political evolution and the failure of the progressive nationalism of Ruiz Jiménez. See Miguel Martínez Cuadrado, *Cambio social y modernización política* (Madrid: Edicusa, 1970), pp. 265–269.

11. In 1957 a National Delegation of Associations was created, and on December 24, 1964, the right of women to form associations was legalized.

12. Celia Valiente Fernández, "Movimientos sociales y Estados: los movimientos feministas en España desde los años sesenta," *Sistema*, 161 (2001): 31–58.

13. Alfonso Ortí, "Jesús Ibáñez, debelador de la catacresis (La sociología crítica como autocrítica de la sociología)," in Jesús Ibáñez, "Sociología crítica de la cotidianidad urbana,"*Anthropos*, 113 (1990): 1–97; and "De la guerra civil a la transición democrática: resurgimiento y reinstitucionalización de la sociología en España" (Inaugural Conference I Congreso FASEE), Zaragoza, Asociación Aragonesa de Sociología, 1982, and Martínez Cuadrado, *Cambio social y modernización política*, p. 267.

14. Martínez Cuadrado, *Cambio social y modernización política*, pp. 267–286.

15. A movement that first emerged in Spain at the end of the nineteenth century which was informed by a pessimistic vision of Spain. It denounced Spain's backwardness and called for the recovery of modernity and the incorporation of Spain into Europe by means of social engineering. Among its principal advocates was the intellectual and writer Joaquín Costa (1846–1911). The teachers who started their careers in the neo-regenerationist period of the 1950s were expected to educate the peasant population in order to elevate the cultural level of a nation that was preparing itself for the social changes necessary to open the way to modernity.

16. Carlos Lerena, "El oficio de maestro. La posición y el profesorado de primera enseñanza en España," *Sistema*, 50–51 (1982): 79–102.

17. Pierre Bourdieu defines *habitus* as a system of corporal and cognitive predispositions, which gathers past experiences and operates like a compass, structures perceptions and generates actions. See Pierre Bourdieu, *El sentido práctico* (Madrid: Taurus, 1975), pp. 93–94.

18. In Spain three types of schools exist: public, private, and state-assisted. The public and state-assisted schools are financed and funded by the State, the difference being that in the former the teaching staff is hired based on the results of a state examination, while each state-assisted school has its own criteria for employing teachers. Private schools are financed by private funds and the teaching staff is chosen by each school. Those teachers who worked in public schools had to face the challenges of the rural villages where

they were sent, while those who worked in private schools were located in cities.

19. In the 1950s many schools were coeducational for economic reasons, and the teaching staff was primarily female.

20. With the change in government a new Plan of Study appeared. The Professional Plan of 1936 established the same curriculum for male and female students in the Normal Schools, where coeducation was permitted. Studies lasted four years, and upon graduation teachers would be guaranteed a civil service position. They received a secular education. With the outbreak of the Civil War, these teachers were purged from public schools. If they wanted to work as teachers after the Civil War they had to present a report signed by a priest. In post–Civil War Spain, women teachers were practicing Catholics.

21. Caroline Steedman, "The Mother Made Conscious: The Development of a Primary Pedagogy," *History Workshop Journal*, 20 (1985): 149–163.

22. The union appeared at the end of the 1960s and the early 1970s with the aim of fighting for employment, the professionalization of teaching, and greater job stability.

23. Consuelo Flecha, "La vida de las maestras en España," *Historia de la Educación*, 16 (1997): 199–222.

24. See Ramón Navarro Sandalinas, "El franquismo, la escuela y el maestro," *Historia de la Educación* 8 (1989): 167–180, especially p. 179, and Agustín Escolano Benito, "Discurso ideológico, modernización técnica y pedagogía," *Historia de la Educación*, 8 (1989): 12–13.

25. The OECD and UNESCO advised the Ministry of Education (Ministerio Nacional de Educación) as to the educational plans they needed to attain the required development. The pressures of the World Bank obliged the regimen to increase its educational funding. Educational spending rose spectacularly in Spain from the 1960s to the 1970s.

26. Rosa Sensat (1873–1961) was an important figure in the history of socialism in Spain. Her ideas on pedagogy, which were closely allied to her socialism, inspired a teacher's movement in Cataluña to reform public schools, introduce new pedagogical innovations, and fight on behalf of education. Marta Mata, the current president of the Consejo Escolar del Estado, is the most prominent representative of this movement. See Esther Cortada Andreu, "Rosa Sensat Vilá: Devoción por la Naturaleza," in Cuadernos de Pedagogía, 337 (2004): 23–26

27. Max Weber, *La ética protestante y el espíritu del capitalismo* (Barcelona: Editorial Península, 1985).

28. The polite form of "you."

29. Political movement founded by José Antonio Primo de Rivera in 1933. Its principal ideological views were based on a concept of one, united Spain, the disappearance of political parties, and the official protection of Spanish religious traditions. *Diccionario de la Lengua Española de Real Academia española* (Madrid: Espasa, 1992), 21 edic.

30. *Ley de Ordenación General del Sistema Educativo*, which was implemented in the 1980s.

31. Max Weber, *La ética protestante y el espíritu del capitalismo*, p. 107.

CHAPTER 2

Collegiality and Gender in Elementary School Teachers' Workplace Cultures: A Tale of Two Projects

Elisabeth Richards and Sandra Acker

Various authors have touted the benefits of collaborative teaching endeavors. Arhar, Johnston and Markel,[1] for example, have found that team-teaching increases satisfaction, reduces isolation, and enhances feelings of professional competency. According to Brookhart and Loadman,[2] collaboration exposes teachers to diverse teaching orientations and teaching experiences, and this exposure allows for greater stimulation of new ideas. Lieberman[3] contends that increased communication allows educators to form close and long-lasting relationships. From these and similar investigations, it would seem that the quality of professional life in schools is greatly improved with increased peer interaction and interdependence.[4]

While there are many positives that are said to be gained with collegial sharing, research conducted at the elementary and secondary level shows that the degree to which school staffs put collaborative values into practice varies enormously. Typically, studies done on secondary schools depict staff relationships as "balkanized"[5] and hierarchical. Organizing high schools around subject subcommunities promotes competition among departments for resources.[6] Teacher loyalties become attached to their own group rather than to the school as a whole, creating strong subject boundaries and accentuating subject status differences.[7] While strong subject affiliations may help facilitate collegiality among staff within the same department, they paradoxically reduce teachers' abilities to forge relationships with those in other departments and lead to various forms of exclusion.

The jury is out on elementary school collegiality. One tradition, epitomized by Lortie's classic account in *School-teacher*,[8] identifies individualism as a prime value for elementary school teachers. Individualism and isolation are thought to be historically created and supported by the physical "egg crate" or cellular layout of the typical elementary school.[9] Others[10] regard the immediacy, uncertainty, and complexity of the classroom as demanding a teacher's full attention, leaving little time or energy for socialization with colleagues.

In contrast, some authors report closeness, collegiality, and egalitarianism among teachers in certain elementary schools.[11] For example, Nias[12] characterizes the primary [elementary] school teachers in her British study as committed to values of interdependence, openness, and trust. Troman and Woods[13] depict the primary school culture as "naturally supportive." Explicitly or implicitly, these studies evoke the feminization of the elementary school—both in the sense of numbers (a preponderance of women) and ethos (qualities usually associated with women). Acker,[14] however, argues that the presence of women is not as important as the hegemony of a gendered script; that is, both male and female elementary school teachers will be influenced by societal expectations that women working with children will think and behave in certain ways.

A third strand of literature comes closer to the model for secondary school teachers' cultures, suggesting that for certain teachers at least, fragmentation and conflict replace warmth and collegiality.[15] Collaboration may be confined to certain groups or cliques within a school[16] or artificially imposed from above.[17] Teachers on the margins do not benefit from the collegial norm. For example, in Feuerverger's[18] study of elementary school heritage language teachers, many participants voiced strong feelings of alienation and isolation. Substitute teachers as well as teachers who are part-time or move from school to school (e.g., to give music lessons) may also have difficulty participating in the core teacher culture in a school.[19] Sargent[20] finds men teachers in elementary schools operate in a fog of "socially constructed expectations," in a workplace culture dominated by women. In this third strand, any beneficial effects of feminization on elementary school teacher culture would be outweighed either by the presence of organizational micropolitics and exclusionary processes or the impact of the wider society and its gendered patterns.

Drawing on these various and contradictory arguments about teacher collegiality and gender, the aim of this chapter is to re-visit and extend debates on the topic. Each of us has done a study of elementary school teachers. At first glance the results seem very different. In Acker's ethnographic study of Hillview Primary School in England, there was a culture of caring among the teachers. In contrast, Richards's interviews with core French elementary school teachers in Ontario, Canada, produced many stories of bitterness and exclusion. In the main part of this chapter we describe each study in more depth and interrogate the different findings, aiming for a more comprehensive view of teacher cultures. First, we sketch the historical context of women teachers' work in each setting.

FEMINIZATION AND THE TEACHING PROFESSION

There are both similarities and differences between the two countries considered in this chapter. Canada is a very large country geographically, with a relatively small population, while Britain is small but densely settled, as well as much older. Canada as a country was established by the British North American Act in 1867. Due to linguistic, regional, and religious differences among the founders, responsibility for education was decentralized to the provinces.[21] Without a national ministry of education, Canada's federal government has not been as interventionist in educational matters as Britain's in recent years, although individual provinces, including Ontario, have dictated substantial changes in teachers' work and teacher education.

In both Canada and Britain, as elsewhere, gender has long been a source of divisions in the teaching profession. Women form the majority of the teaching force, and are numerically dominant, especially evident at the elementary/primary level. The circumstances under which this feminization occurred have been noted by Prentice and Theobald[22] as "subject to much regional variation and ideological complexity," and comprise regional, class, religious, ethnic, and linguistic differences. Our discussion that follows captures the highlights of the histories in order to provide background for our more specific discussions of teacher collegiality. We emphasize England and Ontario, where our two studies take place.

Some introductory generalizations can be made. Women are likely to have played a teaching role well before the advent of compulsory schooling and various forms of organized but informal schooling (such as the "dame schools") have been documented in both places. Well-to-do families could hire a governess or from the end of the eighteenth century send their daughters to ladies' academies,[23] distinctions often being made between (private) schooling for the wealthy and schooling for the general population. The employment of women as teachers is tied in with the expansion of education for the population at large; as schooling became compulsory in the nineteenth century, teachers were needed in larger numbers. In turn, the existence of large numbers of teachers extended and improved systems of teacher education. As time went on, governments increasingly depended on schools for the promotion of economic welfare, patriotism, and other social goals, while waves of curriculum reform and extensions of school-leaving ages required better trained and remunerated teachers. Demographic and economic fluctuations have large-scale repercussions for teachers' well-being and the historical accounts show many ups and downs. The efforts of teachers' unions and federations to improve wages and conditions, largely successful over time, read like sagas with moments of victory interwoven with periods of despair.[24]

Britain

The history of teaching in Britain is complicated and signposted with significant acts of legislation and reports of royal commissions.[25] One could go back to medieval times, but most of the sources that discuss the teaching profession begin with the nineteenth century. The two main themes seem to be the gradual extension of educational provision to a wider segment of the population and the changing role of the state. Initially the national state was reluctant to intervene in the realm of education and in the first part of the nineteenth century several voluntary societies (both religious and secular) provided elementary schooling for the working class, while upper middle-class schooling was private and sometimes residential, especially for boys.[26] In 1870, elementary schooling became compulsory for children between five and thirteen. Not until the 1944 Education Act was the principle established of free secondary education for all. Elementary education was then replaced by a primary sector up to age eleven, followed by a differentiated system for secondary education.[27] Although the intent of the legislation was egalitarian, class divisions in state secondary provision survived, with middle-class children more likely to be in grammar schools (academic secondary schools) while the majority of the population attended secondary modern schools.

Grammar and secondary modern schools might be either single-sex or coeducational. From the 1960s, comprehensive secondary schools, which were usually coeducational, began to replace the differentiated system. The primary phase catered for age groups from four or five to eleven. Heavily influenced by what is known as "The Plowden Report," primary teaching was infused with a child-centered approach that held sway in the system for many years, although there was a backlash against it eventually which reached a peak in the early 1990s. From the 1960s and gaining momentum in the 1970s and 1980s, a second-wave feminist influence produced concerns over sexism and later racism in school practices, biased curriculum materials, and unequal career chances for women teachers. But from the late 1980s, the state began to implement a standards-based agenda within which curriculum, assessment, and governance of schools, as well as teacher education, became much more closely monitored and controlled by the central government.

Histories of women's education in Britain usually make a substantial distinction between the experiences of middle- and working-class girls. Pioneering headmistresses had established a network of fee-paying (middle-class) girls' secondary schools in the last part of the nineteenth century. Women who taught in the middle-class girls' schools went to such schools themselves and then to university, while those who staffed the elementary schools came from lower middle-class or upper working-class backgrounds and did not have access to universities. Elementary education was coeducational although it might be gender-differentiated within the curriculum.[28] Stratification in British society was often reflected in teachers' working conditions, with teachers divided by "geography, qualifications, pay, sex and educational location."[29]

After elementary schooling became compulsory in 1870, more teachers were needed, and from 1870 to 1900, the small group of mostly men "certificated" teachers were outnumbered by untrained or partly trained teachers (many of whom were women), who could be paid less. During that period, women moved from being one-half to three-quarters of the teacher labor force.[30] Even as teaching became more secure and professionalized, divisions—especially gendered ones—remained. The main union for elementary school teachers, the National Union of Teachers, was to see groups of women and men withdraw from its ranks in the first quarter of the twentieth century and form single-sex associations over gender issues.[31] A prominent gender issue for teachers was equal pay, as women and men were paid on different salary scales. Despite much campaigning over the years, equal pay for women teachers was not fully implemented until 1961.[32] Another issue was the so-called marriage bar. Local education authorities required women teachers to resign upon marriage in the early twentieth century, although not universally (rural areas needed teachers too much and London was usually an exception).[33] During World War I, married teachers were kept in their posts but after the war the bars were reinstated[34] It took another world war to change social attitudes sufficiently to abolish this practice.[35]

The other major gender issue concerned promotions. It was generally accepted policy to give men an advantage in appointments to management positions except for headships of separate infant schools (i.e., those for children up to age seven). Sometimes in the earlier part of the twentieth century this practice was the conscious policy of local governments.[36] It reflected other patterns of differentiation whereby "women predominantly taught infant, junior and girls' classes and men were concentrated in the boys' classes, particularly the older boys."[37] Whenever schools were reorganized or amalgamated, generally in the direction of creating larger units or coeducation, men tended to be appointed as heads. Because of this tendency, some commentators regard coeducation as having been highly detrimental to women teachers' careers.[38] Even in the 1980s, teachers believed (with good evidence) that men were preferred for primary school headships. When the head teacher was female, it was often taken for granted that the deputy head teacher should be male.[39]

An apprenticeship scheme for training promising students from the age of 13 to be "pupil-teachers" began in 1846; if successful in an examination after five years they might then attend a training college. The first teacher training college opened in 1805 and the first women's training college in 1841.[40] Most of the colleges were single-sex and residential. Gradually the age at which the apprenticeship could begin was raised (to 16 in 1903) and the system itself was phased out in the early part of the twentieth century. At that point, prospective elementary school teachers were expected to go to a training college for two years. From 1902, some colleges were run by local education authorities rather than voluntary societies.[41] According to Gardner, these changes meant that elementary teaching was becoming a more attractive career, incorporating a subsidized form of education that provided social mobility for its practitioners. Another development from 1890 was the "day training college," an institution attached to a university, which was non-residential. By 1900, nearly a quarter of teacher candidates attended the day training institutions, the precursors of university departments of education.[42]

By the 1940s, it was recognized that about 70 percent of all teachers and an even higher proportion of those in elementary schools were women.[43] Of 83 recognized training colleges in 1942, 60 were for women only, 16 for men, and 7 coeducational. Those trained in the colleges usually taught in the elementary schools while secondary teachers went through a different route, getting a university degree and a one-year training qualification. Over time, various schemes were introduced and teacher training lengthened, except in cases of teacher shortage. In the 1960s, the B.Ed. degree was introduced, beginning the transition to an all-graduate profession. Many of the women's colleges had been made coeducational and the era of the residential college was over.[44]

Maguire and Weiner[45] note that in the first half of the twentieth century, English teacher education was feminized. In the 1940s and 1950s almost three-quarters of college of education lecturers were women.[46] From the 1960s, a growing need for teachers led to the expansion of teacher training as well and the overt recruitment of men.[47] After 1974 the colleges were merged with other institutions or became multipurpose colleges of higher education. Men took over instructional and managerial roles that women had filled in single-sex schools. Commenting on this change, Coffey and Delamont[48] lament: "Thus in thirty years two career paths for women were lost, and nothing was done to replace them."

In recent years, teacher education, like education generally, has come under detailed government scrutiny. Despite the location of most teacher education in universities,[49] the curriculum has nevertheless been prescribed and inspected and the role of primary and secondary schools in the training "partnership" expanded.[50] Teaching at primary level is still a feminized profession. In England in 2000, women were 83.8 percent of teachers in primary and nursery schools and 53.7 percent of those in secondary schools[51] and these proportions have been rising.[52] There are still feminist concerns about promotion barriers for women teachers.[53] Of the women who worked in nursery and primary schools in England and Wales in 2000, for example, 7.9 percent are head teachers [principals] compared to 28 percent of the men.[54] Yet gender equity for women teachers has a low profile in educational policy debates. There has been, however, a revival of an old issue about whether boys are underachieving in "feminized" schools, combined with popular rhetoric about the need for more men teachers.[55]

CANADA

While schooling in private homes was common in the first half of the nineteenth century in Canada, by mid-century it had mostly been replaced by one-room schools.[56] These small "common schools" of the early1800s were originally staffed by men, apparently those who due to some disability or old age were unable to work in more physically demanding occupations.[57] Some teachers worked seasonally, often joining their students in spring and fall field labour.[58] Gradual improvements in salaries and living conditions encouraged women into teaching and they were about a quarter of those employed in the Ontario common schools by 1860[59] and over half by 1880.[60]

From the mid-nineteenth century, provinces like Ontario were divided into school districts, some very small, with elected trustees who used funds from local taxes and provincial grants to maintain schools and hire and pay teachers.[61] In the cities, the districts were more bureaucratized. The Ontario School Law of 1871 made schooling compulsory for children between seven and twelve.[62] Between 1850 and 1900, as the system expanded, women's representation among elementary school teachers jumped from 25 percent to 75 percent in Ontario schools.[63]

Various historians have made a connection between feminization and urban bureaucratic school systems, for Ontario and elsewhere. These systems incorporated hierarchies whereby men taught older children and occupied positions as administrators and inspectors, while women taught younger children. The presence of these systems was associated with both a wage differential in favour of men and an increase in the presence of women teachers.[64] Toronto, Ontario's major city, tended to match the dominant pattern; although it feminized early and was mostly coeducational, it was a stratified and gendered system, with men teaching older children, women the younger ones, and men holding the management positions.[65]

Among women, there were further divisions. Women were expected to leave teaching if they married.[66] Although training opportunities were improving with the founding of "normal schools" for teacher education, trustees might go for women without those qualifications because they could pay them less.[67] A practice called "underbidding," common in the late nineteenth and continuing into the twentieth century, made some salaries still lower, as trustees would hire whoever would agree to a reduced wage.[68] In the beginning, women teachers in Toronto in 1906 were paid only a few dollars more annually than "charwomen at the post office," and salaries in rural areas were even lower.[69] A rhetoric around the supposed naturalness of women's inclinations toward "mothering in the classroom" grew up to support the financial bottom line, although some historians argue that the rhetoric followed the feminization of teaching rather than stimulated it.[70] Of course, there was a certain irony in the fact that for so long, "real" mothers were largely excluded from a career in classroom teaching.

For their part, the teachers increasingly believed they were treated like workers although they saw themselves as professionals. Teachers had concerns not only over pay but about hygiene, sanitation, isolation, and the manual work required to maintain heat and cleanliness in the rural schools. Hesitant to identify fully with labour movements, teachers nevertheless formed union-like associations or federations.[71] Ontario (and Quebec) were unusual in North America in forming gendered and religiously segregated teachers' federations. The struggles of the Federation of Women Teachers' Associations of Ontario (FWTAO) to improve working conditions for women teachers make interesting reading.[72]

Even by the middle of the twentieth century, there were salary differentials and different training paths for elementary and secondary teachers.[73] Elementary teachers rarely had university degrees and were trained in the normal schools, soon to be renamed teachers' colleges, while secondary teachers had degrees and attended the Ontario College of Education associated with the University of Toronto. Moreover, most women in each panel were paid less than men. Women were more than 80 percent of the elementary school labour force, although men dominated the administrative positions.[74]

But change was on the way. Local rules against employment of married women as teachers were gradually dropped during and after World War II[75] and by 1957, the principle of equal pay for men and women teachers had been accepted across the province.[76] From the 1960s onward, as in Britain, second-wave feminism had an impact and sex-stereotyping in the curriculum and women's under-representation in management positions were among the concerns expressed. The FWTAO was shifting its attention away from survival issues like salary to questions of women in leadership positions and maternity leave.

By the mid-1980s, gender equity was part of Ontario Ministry of Education policy and women teachers were enrolling in greater numbers for courses leading to principalships.[77] In 1990, the province set a goal of 50 percent women in "positions of added responsibility" in schools by 2000. The subsequent Conservative government, however, dropped this and other equity initiatives. Nevertheless, considerable momentum occurred and it could even be said that school administration, at least in the elementary schools and at the vice-principal level, was becoming feminized. In 1976–1977, women held only 6.9 percent of public school principalships in elementary schools.[78]

By 1991–1992, women's proportion among elementary school principals rose to 23 percent[79] and in 2000–2001, women held 58.4 percent of elementary and 43.8 percent of secondary school principalships and vice-principalships in Ontario. If classroom teachers and administrators are combined, women constitute 73.7 percent of this group in elementary schools and 50.3 percent in secondary schools.[80]

In nineteenth century Ontario, some teachers found training opportunities in "model schools" (schools where students could do practice teaching) but few of these remained by the start of the twentieth century, supplanted by the normal schools, later renamed teachers' colleges. Normal schools were rigid conveyers of the "norms" and highly controlled by provincial departments of education.[81] They were staffed mostly by men. Teachers in secondary schools were expected to have university degrees rather than teacher training and what training there was took place mostly in universities. Over time, the teachers' colleges were absorbed into the universities and became faculties of education and the segregation of elementary and secondary training disappeared[82] Currently in Ontario, almost all teacher training takes place in universities and mostly follows the pattern of degree first, then one year of teacher education. Most of the teacher education was and is coeducational and (probably as a consequence) there does not seem to have been a period of feminization of teacher education equivalent to the one found in England in the first half of the twentieth century.

Summary

In England and Ontario, there were divides on the basis of gender and other characteristics, although the literature appears to portray the divisions as deeper and more intransigent in England, where social class permeated educational provision in exaggerated ways. In both cases, the teaching occupation gradually professionalized and improved its working conditions, often through hard-fought battles between teachers' unions and governments. Coeducation was the norm in Ontario and teacher education non-residential; while England had a history of residential, single-sex training colleges and single-sex schools, providing a different pattern of opportunity (and foreclosure of opportunity after coeducation became the norm). Struggles for equal pay and for the rights of married women to continue employment as teachers occupied teachers' unions in both places.

The studies to be described below concern teachers in what in England would be called the primary phase and in Ontario the elementary one. Currently, it appears that English primary schools are slightly more "feminized" than their Ontario counterparts, judging from the statistics cited earlier.[83] In both sites, there have been expressions of concern over what appears to be a recent trend toward the further feminization of primary and secondary school teaching.[84] We will return to some of the contextual differences later in the chapter. For now, we note that there is no obvious clue in the histories of the two settings that would lead us to predict differential levels of collegiality among primary school teachers. A discourse about the mother-like qualities of teachers of young children is widespread both in Britain and Canada. One might argue that a proclivity for "caring" would extend into close relationships with colleagues; but equally one could say that the classroom teacher would focus her nurturing on her students to the exclusion of other adults. Let us now turn to our empirical evidence on the question.

Are Elementary Schools Collegial?

Yes: The Acker Study

From April 1987 to December 1990, Sandra Acker conducted the major part of an ethnographic study of Hillview Primary School, reported in detail in her book, *The Realities of Teachers' Work: Never a Dull Moment*.[85] Hillview served the primary age range from age four to age eleven and contained about 200 children. Throughout most of the research the staff included a head teacher [principal], a deputy head teacher, and eight others, three of whom were part-time teachers. All of the teachers were women except for Dennis,[86] the deputy head [vice-principal] and class teacher of the oldest children in the school. All but one were currently married, the other divorced. All but two had children. Dennis was the only non-white teacher. All teachers were born in England or Wales. Six staff were in their late twenties or thirties, and four in their forties or fifties.

Gender and Caring in the Classroom
The study as a whole concentrated on the themes of teaching as work, gender, and workplace culture. Hillview displayed many qualities of what might be thought of as a women's culture. Not only were most of the staff women, there was much evidence of caring, kindness, consideration, and connectedness in their interactions.

Acker's book includes analysis of the contradictions involved in enacting the caring script. While there are numerous instances that depict teachers as deriving an enormous amount of satisfaction from their work, Acker also shows the great difficulties and frustrations embedded in working for many hours with large numbers of children of diverse abilities and motivational levels. These frustrations were compounded by deficiencies in the workplace: an old and cramped building, an inner-city site without adjacent playing fields and a serious shortage of resources. Certainly, these teachers cared deeply for their students, but the ability to give care was something that required conscious effort. To cope with the stresses of such a daunting task, teachers often turned to colleagues for support.

Collegiality at Hillview

Hillview's teachers' workplace culture featured participation and equality; caring and mutual support; and a sense of community.

Teachers spent a great deal of time meeting and talking together. Staff meetings were frequent (once a week) and characterized by a high level of participation in decision making. There were extended considerations of arrangements for assemblies and productions, resources, dates, projects, visits, evenings for parents, schedules for swimming or games, as well as discussion of curriculum or school policy topics such as handwriting, multicultural education, spelling, science, safety. One teacher's comment "Shall we talk about Christmas?" ushered in lengthy discussions, carrying on over several months leading to the actual celebrations. All teachers freely contributed in follow-up sessions, and plans, such as singing carols and making sheep costumes, were created collectively. A belief in the importance of on-going communication was also demonstrated by consultation and joint planning among subgroups of teachers. Teachers saw one another before and after school, during morning assembly, at morning playtime (coffee time for the teachers), and at lunchtime. Whether their interactions were planned or spontaneous, the teachers took pride in their ability to work as a team.

Most striking was the level of concern and compassion teachers showed *for one another*. Rosalind, a part-time teacher, was about to leave for the day, but stepped in to take Sheila's class when a family emergency arose. Betty volunteered to take over the organization of the library when Sheila went on maternity leave. In an interview, Marjorie talked about an episode from a few years back when she had volunteered to change places with another teacher having difficulty with a "tough class." When one teacher went out to relieve another on playground duty so she could come in for coffee and cake, Joan Miller, a retired teacher who did some volunteer work in the school, remarked to the researcher: "How wonderful it is to see teachers here caring about each other that way. It doesn't happen everywhere."

The head teacher took the lead in showing care and concern for her staff. She gave teachers encouragement, praise, and gifts such as chocolates after a Christmas production. She offered individuals career advice, steered them toward useful courses and higher degrees, and took into consideration their domestic and child care responsibilities in making plans for the school.

At Hillview the staffroom was a place where solidarity flourished. An example from field notes conveys the sense of togetherness:

> [On a day near the end of term] scones and tea are in the staffroom. People are coming in with coats. There is a jovial atmosphere. There are two bottles of wine there, for Lisa's [a student teacher] last day. Kristin has brought in scones and jam for her birthday on Sunday. Everyone sang to her.

Some of the teachers socialized together outside of school. This interaction was especially apparent for Debbie, Rosalind, Kristin, Sheila, and Dennis, all teachers with young children of their own. This bond, as well as the authority position he held as deputy head teacher, seemed to counter any possibility that Dennis as the only Black and only male teacher might be marginalized from the workplace culture.

Supports for the Collegial Culture

What appears to have happened at Hillview is that teachers turned to one another for support in what was often a challenging teaching context. Teachers taught under difficult circumstances that left them feeling angry and frustrated. These circumstances pre-dated the additional stresses usually blamed on British government policy from 1988 onward.

Some of the problems stemmed from high expectations teachers held for themselves, quite possibly derived in part from widespread beliefs about *women's* work and its "labour of love" nature—set against material conditions that worked against high standards. The classes were large; the resources were poor; and the building created difficulties of noise and movement. The community served by the school was very diverse and included a number of unconventional or "alternative" families. The school ethos was that children's differences were to be welcomed; some eccentricity should be tolerated; children should be academic but not without a social conscience. To work in this school required teachers to have considerable skill, patience, and tolerance.

This combination of factors appeared to give a strong impetus to the development of a caring, mutually supportive teachers' workplace culture. The culture might be thought of as feminized: there was a preference for sharing rather than demarcating responsibilities; participation over delegation; equality over hierarchy; support over competition. Humor was integrative rather than divisive. Community feeling was sustained by rituals. This school was certainly collegial.

No: The Richards Study

The goal of Elisabeth Richards' study was to interrogate assertions of collegiality by examining how subject hierarchy is conceived by those who are situated on the fringe of elementary education. Richards interviewed 21 teachers who taught "core French" in elementary schools in Ontario, the most populous province in Canada.

In Ontario, "core French" usually begins in Grade 4 (age 9), although local school boards can introduce the subject earlier. It is a mandatory subject, provided in 20–40 minute daily segments until the end of Grade 9 (age 14), when it becomes optional.[87] According to the 1998 core French curriculum document, students must have accumulated 600 hours of French instruction by Grade 8 (age 13).[88]

The interviews were conducted between the fall of 1997 and the spring of 2001 and ranged from one to one and a half hours. While the sample was not large and was relatively homogeneous (the sample consisted mainly of white, female teachers between the ages of thirty-five and fifty who spoke French as a second language), the interviews were in-depth and allowed a normally silent group to express their views.

Elementary Core French as a Marginal Subject
Elementary core French teachers can be seen as a marginalized group in comparison to regular classroom teachers who constitute the norm in elementary schools. Classroom teachers, sometimes called home-room teachers in North America, have charge of most of a child's elementary school education, especially the "basic" subjects such as English language and mathematics. Typically, the school class becomes a tightly knit social group throughout the school year, staying together for most of its learning experiences. The class teacher is closely identified with "her class."[89]

In contrast to the class teacher, the core French teacher is a visitor, often without a room of her own, teaching approximately 200 students a day and moving from class to class.[90] Moreover, while boards of education receive substantial transfer payments from the Federal government of Canada to fund French programs, seldom do core French teachers see this money reserved specifically for their use.

Ideologically, this unequal distribution of rewards is justified by a rhetoric that serves to present these divisions as perfectly "natural." The subjects taught by homeroom teachers, for example, are the ones identified for province-wide testing. Others—art, music, French, physical education—appear by contrast as "frills."

By unequally privileging some groups of teachers over others in terms of resources (e.g., amount of physical space, budgets, numbers of students) and symbolic prestige, it can be argued that schools work to create a mythical center which structurally marginalizes, among others, elementary core French teachers. The question Richards' study sought to address was do elementary core French teachers discover this marginality, or are there counter-discourses of inclusivity operating in the elementary school system that would mitigate against the sense of disenfranchisement that such teachers might experience?

Responding to "Difference"
There were times when the participants positioned themselves as having the same value or freedom of action as other teachers in the school. Margot, for example, described herself as "one of the crowd," while Farrah believed that "people don't see me as any different than anybody else." Participants experienced themselves as being a "real" teacher when they felt their authority was taken seriously. Jennifer, for instance, noted, "I think that my opinions are listened to by a lot of the staff," while Theresa commented, "I have found whenever I have a problem with a particular student and I send them down to the principal, I usually never have that problem with that child again."

The majority of participant commentary expressed exclusion from the definition of a "real" teacher. The understanding of a barrier existing between the homeroom and the elementary core French teacher related to the system whereby the homeroom teacher had "time off" while the core French teacher taught her class:

We're kind of the babysitters. . . . We'll cover this class. (Katya)
We're the spare [free time] providers. (Roxanna)
You are just a place to go for awhile. (Ken)
Sometimes I see myself as a walking coffee cup. (Julie)
[Core French teaching] is conceived to give people a break. . . . I know that in my being, I am a first-hand teacher, but in fact, if I look around, I am a second-hand teacher. (Greg)

Because core French teachers often felt that their main function was "to give homeroom teachers a breather" (Samantha), they tended to describe themselves as an extension of someone else's being. Rachel, for example, said, "I just felt like an appendage really. My perspectives were not necessarily respected or considered," echoed by Natasha, "It's sort of like you're not a real part of the team, like a little appendage or something."

Female teachers discussed their gender as being a catalyst for violence that further complicated the relationships of power and authority operating within the core French classroom. Brenda, for example, described a Grade 8 [13 year olds] core French classroom where her gender was used as an object of attack: "They [the students] just made it miserable, changing their names so I couldn't figure out who was who, throwing things, just aggressive, violent, really mean spirited. . . . I was told to fuck off and called stupid bitch constantly." A female core French teacher's legitimacy could be further compromised by her race:

> [Students can be] physically threatening, . . . they can come up very close to me, or be verbally threatening. Just the other day one said, "Why don't you go back to where you came from?"(Rachel)

While the harassing gestures of the students getting "too close" remind Rachel of her gendered status, racial slurs compound her positioning.

Though the complex interplay of core French teaching identities within racialized and gendered discourse demonstrates that marginality is not uniform in its causes or effects, it is still the case that the majority of participants did not find themselves located in school cultures they could describe as collaborative. In fact, there was evidence in Richards' study that certain forms of teacher collaboration could only be achieved through the implicit or explicit exclusion of "the other." Roxanna, for example, spoke of providing a spare [a free period] so that homeroom teachers could plan together, while Julie talked about the necessity of the elementary core French teacher acting as the in-school supply [substitute teacher] when homeroom teachers went on school trips.

Some of the participants were suspicious of the rhetoric around "caring" in the elementary school. Alexandra, for instance, said, "I think there's a real problem with thinking about elementary teachers as all being on the same team. It makes the elementary core French teacher feel like they can't complain because then they won't be acting as a team player." Like Roxanna and Julie, Alexandra experiences this form of collegiality as *mythological collaboration*. Hargreaves[91] describes contrived collegiality as a covert way of increasing control over staff members and to inveigle them into complying with administrative agendas. Yet, from these interviews, it appears that this need to cajole the staff into acquiescence is not necessary for core French teachers, as they are already disciplined into submission by virtue of their marginalization. In other words, it is through the illusion of "sameness" among teachers that various inequalities or contradictions within the system are obscured. In this sense, the collegiality that core French teachers experience is not contrived, but rather mythological[92]— something they provide for others but never actually participate in themselves. The elementary school may be a "home" for "homeroom" teachers, while enforcing forms of exclusion for members of marginal teaching populations.

Discussion

As is evident from the above reports of the Acker and Richards studies, we reside in different strands within the literature on elementary school teacher culture. Acker's ethnography is consistent with the research identifying elementary schools as collegial and caring places, while Richards' interviews produce evidence for the operation of dividing practices and consequent marginalization within the teacher population.

What might account for the differences in our results? We offer four explanations as to why schools may be depicted as more or less inclusive. They are as follows: (1) the standpoint of the researched; (2) the standpoint of the researcher; (3) the approach of the study; (4) the location of the study.

The Standpoint of the Researched

Standpoint theorists have argued that one's social situation provides the basis for what one can know.[93] Because those with more social power (such as men in a male-dominated society) are invested in preserving a system that serves their interests, they are less likely to generate critiques of the status quo. Acker's study focused on the administration and the core staff of Hillview. Each of these teachers (apart from the head teacher/principal) had a classroom base, although some classrooms were shared between two and in one case three teachers. The teachers' culture stemmed from several factors, including the head teacher's values, the school-wide ethos that emphasized tolerance and diversity, and the effort to cope with difficult circumstances through standing together and sharing strategies. In contrast, Richards intentionally singled out teachers who occupied what appeared to be a structurally marginalized position. The implication of her findings is that a primary teacher's sense of belonging may have more to do with the rewards they obtain from being in that system than with the school's workplace culture per se. Interestingly, a reference to the peripatetic [itinerant] violin teacher in *The Realities of Teachers' Work* appears to support Richards' analysis, when Acker mentions that the violin teacher "risked missing an announcement or a cup of tea because no one remembered she was there."[94]

The Standpoint of the Researcher

It is not only the researched but also the researcher who is subject to the impact of positioning. Acker's ethnography was conducted in mid-career, after years as a university teacher in a School of Education. Her prior work had been concerned with women and education. Although trained as a secondary teacher, she had limited first-hand experience teaching in the schools, although she had many years of instructing teachers, as well as personal experience of growing up as the daughter of teachers and being a parent of a student in school. Richards' study was a doctoral thesis, representing an early stage in a potential academic career and following a number of years working as a core French elementary school teacher.

In her article "In/out/side: Positioning the researcher in feminist qualitative research," Acker[95] discusses how various combinations of insider and outsider status vis-à-vis the community studied affect both acceptance and perceptions. In the Hillview research, Acker came from outside the school setting and did not fully join it, attempting to retain the perspective of the sociologist. In contrast, Richards hailed from the group under analysis, but did her study as a doctoral student and had moved toward the outskirts of the group. Richards acted as a standard-bearer for core French elementary school teachers, but her interests as a researcher were also in theorizing marginality, not simply in improving teaching conditions.

Acker was amazed to see how hard the teachers worked and how hard the work itself was. Her background in gender analysis led her to connect the lack of public appreciation for teachers with the female dominance of the field, numerically speaking. She was very motivated to show that the teachers were hard-working professionals who were, nevertheless, influenced by a social script that likened teaching to mothering. The high level of collegiality in the school was a *finding*; her interpretation of it was that the difficult conditions under which teachers worked contributed to that kind of collegiality developing as a solution.

Acker chose the main set of teachers and the head teacher as her core group. Others, such as the peripatetic music teachers, or personnel from the local authority [school district] such as the nurse or educational psychologist, or the student or substitute teachers, came into the observational part of the study and are quoted from field notes, but they were not "central" and thus were not interviewed.

From Richards' perspective, Acker could not get a full picture by concentrating on core staff. Richards' approach was undoubtedly influenced by her prior experience of working in a marginalized subject area. Since she knew the dominant homeroom teaching culture from the perspective of someone who largely has been excluded from it, her cultural understandings of collegiality had evolved in isolation from those of the homeroom teacher. Certainly, Richards' interviews confirmed her suspicion that there were teachers on the fringes of the school who have been largely left out of mainstream research on teachers' work.

We see how interpretations can never be firmly anchored to a Euclidian space that somehow lies outside of the researcher's assumptions and predispositions. Yet, we can go beyond our prior experience and understandings; if we could not, there would be little point in doing research.

The Approach of the Study

Each of the researchers depended on a different main research method. Following standard techniques for ethnography, Acker spent many hours in the classrooms with the teachers and frequented other places where they gathered such as the staff room, the playground, the head teacher's office, and various staff members' houses. She went on trips to the zoo, on the regular swimming and games expeditions, to school assemblies, and many other places. She also collected many documents, such as curriculum documents issued by the government and school-produced letters to parents. Her other main method was the in-depth interview. Some teachers and the head teachers were interviewed a number of times.

Richards prioritized the interview and worked with teachers who shared a teaching subject but not a school. While Acker's teachers might be hesitant to say anything critical about other teachers (although in practice, they seemed remarkably open, perhaps because of the lengthy period over which trust was gained), Richards' subjects would have no comparable concerns about their words getting back to close colleagues, although given the content of their interviews, they might well have worried about anything being reported that could identify them. Acker herself may have been careful to avoid commentary that appeared critical of individuals, given the level of trust placed in her by the school staff and the likelihood that each participant would read what was said about herself and others known to her. Richards would not have had the same concern as her participants were not based in one place.

The Location of the Study

A fourth suggestion might be that certain socio-historic configurations allow for greater teacher cohesion than do others. Whereas studies done in Britain tend to describe the primary school climate as collegial, the North American literature presents a more mixed picture. The question of why British primary teachers are viewed as having a more collectivist orientation than their North American counterparts is a complex one. Our historical discussion does not immediately answer this question. It may be that there are deep cultural differences, akin to those identified by Ralph Turner[96] when he contrasted "sponsored" (British) with "contest" (United States) norms. The United States is identified with capitalism and individualism; while Britain has a welfare state, state-provided medicine, almost exclusively public universities and other forms of communal commitment. Canada is, as always, somewhere in between.

Some of the cross-national differences in school cultures may be relatively simple to understand, related to structure and traditional organizational forms. For example, urban schools in Britain tend to be smaller than those in North America. Hillview, with 200 children, one head teacher [principal], and 9 teachers (3 of whom were part-time) is considered an average size school. The largely collaborative schools studied by Nias et al.[97] in Britain had rolls of 155 to 262 children and teaching staffs numbering less than nine full-time equivalents. In contrast, Vista City Elementary, in Biklen's[98] U.S. study, had 800 students and 60 instructional staff. Many specialists such as a library teacher, gym teachers, resource teachers, an instructional specialist, a special education aide, and others are mentioned in Biklen's book. No comparable specialists existed at Hillview. Unlike Ontario where teachers are sometimes designated to teach one specific subject, all of the Hillview teachers were classroom (homeroom) teachers, expected to teach across the curriculum. Hillview's head teacher [principal] taught occasionally and the deputy head [vice-principal] had a full class teaching load. A few visiting teachers (violin, guitar, remedial help, multiculturalism) had specific roles but nothing equivalent to the daily input of the core French teacher described by Richards. None was even near full-time in the school.

Another key difference concerns the age of the children. In many of the Ontario schools, "Grade 7 and 8" were part of elementary schools, meaning that the core French teachers had to be able to discipline children who were 13 or 14 years of age. British primary schools rarely went beyond age 11 or 12, similar to the K-6 structure in North America (kindergarten through Grade 6). Given the greater tendency of elementary schools in Ontario toward subject specialization and the older ages of the students, we might expect their character to be more divisive than the elementary schools in Britain and thus to more closely emulate the "balkanized" climate of secondary schools.

It is important to appreciate these national differences. Without a focus on national filters, we are in danger of subscribing to an essentialist model of "the teacher" or "teachers' work" that mysteriously transcends national boundaries. Even in the age of globalization, this is unlikely to be the case.

CONCLUSION

In comparing our assumptions and results, we realized that both of us tended to treat collegiality and collaboration as equivalent to empowerment. Revisiting these concepts and the literature in light of our findings suggested to us that there were other possibilities. We identified three situations where the goal of collegiality conflicts with the goal of empowerment.

The first case is when collaboration is imposed upon teachers. Hargreaves'[99] argument that imposed collegiality becomes "contrived" is now well known. In a more recent study of teachers in Britain several years after the introduction of a National Curriculum and other extensive government-initiated reforms, Osborn, McNess, and Broadfoot[100] add another twist on the idea of imposition, in that teachers might "choose" to collaborate in order to cope with a difficult situation, in this case an overloaded curriculum and the consequent stress. Although teachers in their study said positive things about collaboration, they also regretted the loss of informal and spontaneous exchanges and the sharing with other teachers they had enjoyed in the past (much like the Hillview teachers' culture). They disliked so much of their time being structured and spoken for.[101] Osborn et al. argue that recent studies of teachers in several countries show that "loss of control is linked to a sense of loss of identity and sense of self. It is when teachers no longer feel in control of their working lives that collegiality can become a destructive, rather than potentially a creative, force."[102]

The second impediment to empowerment concerns the impact on individuals of the conformity that frequently accompanies collaboration and collegiality. Hargreaves[103] mentions the possibility of a teacher who prefers to work alone risking ostracism. We think that the consequences may extend further. For example, a teacher may have to compromise her political stance in order to be accepted by colleagues. She may need to hide a facet of her identity, such as sexual orientation, that might not be welcomed by others.[104] Or, like some of the elementary core French teachers, she may feel she cannot complain about perceived injustices because that might disrupt staff "harmony."

Third, collegiality has different consequences for individuals in different structural positions: the message of the Richards study and one that is rarely taken up in the literature. For the elementary core French teachers, supporting colleagues meant denying their own need. Their experience was one of *mythological collegiality*. It seems that a close-knit group of core teachers invites exclusion of those who are not available for the continuous interaction that sustains that subculture. Even in Acker's study, the three part-time teachers faced barriers to full participation. Although they were well integrated into the teacher culture, they were constantly scrambling to find out what they had missed and beset by a sense that something important had gone on in their absence. Unfortunately, most work done in the area of collegiality either excludes certain types of teachers from study or fails to specify the power differences which underpin these understandings of collegiality.

We have come to the conclusion that neither of us holds an exclusively correct version of staff relationships in elementary schools. While Acker emphasizes the possibility of celebratory collegiality, Richards offers a vision in which collegiality is at best, intermittent, and at worst, an ideological contrivance which averts critical eyes from seeing the inequity of systemic disenfranchisement. Most of the teachers in both studies were women. Although there has been an extensive history of feminization in the elementary schools in both

England and Ontario, there has also been a history of male control and struggles for women to gain equal rights. Even in contemporary times, when many (although not all) of the battles have been won, there is no guarantee that women teachers will find themselves in a congenial, empowering, stimulating workplace. Thus, feminization may be a necessary but not a sufficient precondition for collegiality.

Looking at both of these projects simultaneously, one can see that there is room for a third answer to the question "are elementary schools collegial?": one in which elementary teachers are neither heroes nor villains but something in-between. Homeroom teachers may have the desire to help elementary core French teachers but are so overwhelmed by their own workload that they cannot enact their vision of caring. Or, an administrator may support a core French teacher, but may not understand her difficulties as systemic rather than personal. This alternative perspective offers a way out of an ideological stalemate, creating a discursive space where more rather than less dialogue is encouraged. Collegial cultures are possible, under certain circumstances, and not necessarily for everyone in the school. The "certain circumstances" present the challenge of accumulating carefully planned case studies until we have more understanding of what situations are most conducive to improved working environments for all teachers.

NOTES

The authors would like to thank the participants in both studies as well as Linda Button, Mika Damianos and Elizabeth Smyth for assistance with the historical review.

1. Joanne Arhar, J. Howard Johnston, and Glenn Markel, "The Effects of Teaming and Other Collaborative Arrangements," *Middle School Journal* 19 (1988): 22–25.
2. Susan Brookhart and Willliam Loadman, "School–University Collaboration: Different Workplace Cultures," *Contemporary Education* 61 (1990): 125–128.
3. Ann Lieberman, "The Meaning of Scholarly Activity and the Building of Community," *Educational Researcher* 21 (1992): 5–12.
4. Judith Warren Little and Tom Bird, "Instructional Leadership 'Close to the Classroom' in Secondary Schools," in W. Greenfield (ed.), *Instructional Leadership: Issues, Concepts and Controversies* (Boston, MA: Allyn & Bacon, 1987), pp. 118–138; Geoff Troman and Peter Woods, *Primary Teachers' Stress* (New York: RoutledgeFalmer, 2001).
5. Andy Hargreaves, *Changing Teachers, Changing Times: Teachers' Work and Culture in the Postmodern Age* (London: Cassell, 1994).

6. Stephen Ball, *The Micro-Politics of the School* (London: Methuen, 1987); Ivor Goodson, *The Making of Curriculum* (Lewes: Falmer Press, 1988).

7. Andy Hargreaves and Robert Macmillan, "The Balkanization of Secondary School Teaching," in Leslie Siskin and Judith Warren Little (eds.), *The Subjects in Question* (New York: Teachers College Press, 1995), pp. 141–171.

8. Dan Lortie, *School-teacher: A Sociological Study* (Chicago, IL: University of Chicago Press, 1975).

9. Ibid.

10. Sharon Feiman-Nemser and Robert Floden, "The Cultures of Teaching," in Merlin Wittrock (ed.), *Third Handbook of Research on Teaching* (London: Macmillan, 1986), pp. 505–526.

11. Sandra Acker, *The Realities of Teachers' Work* (London: Cassell, 1989); Hargreaves, *Changing Teachers*; Jennifer Nias, Geoff Southworth, and Robin Yeomans, *Staff Relationships in the Primary School* (London: Cassell, 1989).

12. Jennifer Nias, Primary *Teachers Talking: A Study of Teaching as Work* (London: Routledge, 1989).

13. Troman and Woods, *Primary Teachers' Stress*, p. 129.

14. Acker, *Realities*.

15. Grace Feuerverger, "On the Edges of the Map: A Study of Heritage Language Teachers in Toronto," *Teaching and Teacher Education*, 13, 1 (1997): 39–53; Cindy Lam, "The Green Teacher," in Dennis Thiessen, Nina Bascia, and Ivor Goodson (eds.), *Making a Difference about Difference: The Lives and Careers of Racial Minority Teachers* (Toronto: Garamond Press, 1996), pp. 15–49; Elisabeth Richards, *Positioning the Elementary Core French Teacher: An Investigation of Workplace Marginality* (Doctoral diss., the Ontario Institute for Studies in Education of the University of Toronto, 2002); Paul Sargent, *Real Men or Real Teachers? Contradictions in the Lives of Men Elementary School Teachers* (Harriman, TN: Men's Studies Press, 2001).

16. Sari Knopp Biklen, *School Work: Gender and the Cultural Construction of Teaching* (New York: Teachers College Press, 1995).

17. Hargreaves, *Changing Teachers*.

18. Feuerverger, "On the Edges of the Map." Heritage language is a language program usually held after school hours. Much like the core French program to be discussed in this chapter, its goal is to enhance the students' understanding and appreciation of a second language.

19. Beth Young and Kathy Grieve, "Changing Employment Practices? Teachers and Principals Discuss Some 'Part-time' Employment Arrangements for Alberta Teachers," *Canadian Journal of Educational Administration and Policy*, 8 (1996): 1–14.

20. Sargent, *Real Men*, p. 162.

21. Elizabeth M. Smyth, " 'It Should be the Centre . . . of Professional Training in Education': The Faculty of Education at the University of Toronto: 1871–1996," *Tidskrift för Lärarutbildning och Forskning (Journal of Research in Teacher Education*, Umea, Sweden), special issue 3–4 (2003), 135–152.

22. Alison Prentice and Marjorie Theobald, "The Historiography of Women Teachers: A Retrospect," in Alison Prentice and Marjorie Theobald (eds.), *Women Who Taught: Perspectives on the History of Women and Teaching* (Toronto: University of Toronto Press, 1991), pp. 3–33.

23. Ibid, p. 9.

24. Mary Labatt, *Always a Journey: A History of the Federation of Women Teachers' Associations of Ontario 1918–1993* (Toronto: FWTAO, 1993).

25. A good overview text on the history of English educational policy is Peter Gordon, Richard Aldrich, and Dennis Dean, *Education and Policy in England in the Twentieth Century* (London: The Woburn Press, 1991).

26. Gordon, Aldrich, and Dean; Felicity Hunt, "Introduction," in Felicity Hunt (ed.), *Lessons for Life: The Schooling of Girls and Women 1850–1950* (Oxford: Blackwell, 1987), pp. xi–xxv.

27. Hunt, "Introduction," p. xix.

28. Ibid., p. xiv.

29. Martin Lawn, *Servants of the State: The Contested Control of Teaching 1900–1930* (London: The Falmer Press, 1987).

30. Ibid., p. 7.

31. See Lawn, *Servants of the State*; Margaret Littlewood, "The 'Wise Married Woman' and the Teaching Unions," in Hilary de Lyon and Frances Migniuolo (eds.), *Women Teachers: Issues and Experiences* (Milton Keynes: Open University Press, 1989), pp. 180–190; Alison Oram, "Inequalities in the Teaching Profession: The Effect on Teachers and Pupils, 1910–39," in Felicity Hunt (ed.), *Lessons for Life: The Schooling of Girls and Women 1850–1950* (Oxford: Blackwell, 1987), pp. 101–123.

32. Jane Miller, *School for Women* (London: Virago Press, 1996).

33. Lawn, *Servants of the State*.

34. Gordon, Aldrich and Dean, *Education and Policy*, p. 131.

35. Ibid., p. 134.

36. Oram, *Lessons for Life*, 115.

37. Ibid., 120–121.

38. Amanda Coffey and Sara Delamont, *Feminism and the Classroom Teacher* (London: RoutledgeFalmer, 2000).

39. Acker, *Realities*.

40. Elizabeth Edwards, *Women in Teacher Training Colleges, 1900–1960: A Culture of Femininity* (London: Routledge, 2001), p. 5.

41. Philip Gardner, "Reconstructing the Classroom Teacher, 1903–1945," in Ian Grosvenor, Martin Lawn, and Kate Rousmaniere (eds.), *Silences and Images: The Social History of the Classroom* (New York: Peter Lang, 1999), pp. 123–144, 130.
42. Edwards, *Women in Teacher Training Colleges*, pp. 6–7.
43. Ibid., 13.
44. Ibid.
45. Meg Maguire and Gaby Weiner, "The Place of Women in Teacher Education: Discourses of Power," *Educational Review* 46, 2 (1994): 121–139.
46. Ibid., 126.
47. Ibid., 128.
48. Coffey and Delamont, *Feminism and the Classroom Teacher*, p. 113.
49. Former polytechnics, some of which had incorporated colleges of education, were restructured as universities in 1992.
50. Pat Mahony and Ian Hextall, *Reconstructing Teaching* (London: RoutledgeFalmer, 2000).
51. Department for Education and Skills, *Statistics of Education*, 2002, p. 51. There has been a slow move toward (mild) feminization of the secondary school teaching force. In 1985, men comprised 54 percent of secondary school teachers in England and Wales. See Sandra Acker, "Rethinking Teachers' Careers," in Sandra Acker (ed.), *Teachers, Gender and Careers* (Lewes: The Falmer Press, 1989), pp. 10–11.
52. Department for Education and Employment, *Statistics of Education*, 2000.
53. Acker, *Realities*; Coffey and Delamont, *Feminism and the Classroom Teacher*.
54. Department for Education and Skills, 71.
55. Mahony and Hextall, *Reconstructing Teaching*.
56. Alison Prentice, "From Household to School House: The Emergence of the Teacher as Servant of the State," in Ruby Heap and Alison Prentice (eds.), *Gender and Education in Ontario* (Toronto: Canadian Scholars' Press, 1991), pp. 7, 25–46.
57. John George Althouse, *The Ontario Teacher: A Historical Account of Progress*, 1800–1910 (Toronto: Ontario Teachers' Federation, 1967, orig. 1929).
58. Ibid., p. 18.
59. Ibid., p. 48.
60. Ibid., p. 79.
61. Pat Staton and Beth Light, *Speak with Their Own Voices* (Toronto: Federation of Women Teachers' Associations of Ontario, 1987), p. 7.
62. Marta Danylewycz and Alison Prentice, "Teachers' Work: Changing Patterns and Perceptions in the Emerging School Systems of Nineteenth- and Early Twentieth-Century Central Canada," in Alison Prentice and Marjorie Theobald (eds.), *Women Who Taught: Perspectives*

on the History of Women and Teaching (Toronto: University of Toronto Press, 1991), pp. 136–159, 140.

63. Robert Stamp, *The Schools of Ontario, 1876–1976* (Toronto: University of Toronto Press, 1982).

64. Prentice and Theobald, *Women Who Taught*, p. 5. Yet even here there are exceptions: Quebec (the main French-speaking province in Canada) had a different history, which saw women becoming the majority of rural teachers early on, as well as members of religious teaching orders in Catholic convent schools, while men dominated the urban bureaucratized system in the Quebec city of Montreal (Prentice, "From Household to School House").

65. Prentice, "From Household to School House."

66. Labatt, *Always a Journey*, p. 30.

67. Staton and Light, *Speak with Their Own Voices*, p. 21.

68. Labatt, *Always a Journey*, p. 27; Stamp, *The School of Ontario*, p. 112.

69. Stamp, *The School of Ontario*, p. 75.

70. Prentice and Theobald, *Women Who Taught*, p. 6.

71. Danylewycz and Prentice, "Teachers' Work," p. 153.

72. Labatt, *Always a Journey*; Staton and Light, *Speak with Their Own Voices*. Ontario has a Catholic or "separate" school board alongside the secular one, and taxpayers can opt for their school taxes to go to one or the other. Another small board covers schools whose language of instruction is French. Teachers' federations divide into elementary and secondary, secular or separate. In 1998, the FWTAO merged with another federation to become a coeducational union for public elementary school teachers. See Robert Gidney, *From Hope to Harris: The Reshaping of Ontario's Schools* (Toronto: University of Toronto Press, 1999).

73. Gidney, *From Hope to Harris*, p. 20.

74. Ibid., p. 21.

75. Staton and Light, *Speak with Their Own Voices*, p. 143.

76. Ibid., p. 137.

77. Gidney, *From Hope to Harris*, p. 163.

78. Labatt, *Always a Journey*, p. 238.

79. Ibid., p. 297.

80. Ontario Ministry of Education, 2003. These figures give the proportion of administrators who are women, but are not presented in a way that allows calculation of the proportion of women who hold principalships (the figure given earlier for England). Also there is a note that the statistics are for full-time-equivalent teachers and administrators, whereas statistics for previous years in Ontario used head counts. Changes like this one and different conventions from one jurisdiction to another make the use of comparative statistics risky. For example, figures in North America usually separate classroom teachers from administrators, while in Britain they are often combined.

81. Stamp, *The School of Ontario*, p. 201.

82. See Sandra Acker, "Canadian Teacher Educators in Time and Place," *Tidskrift för Lärarutbildning och Forskning (Journal of Research in Teacher Education*, Umea, Sweden), special issue 3–4, (2003): 69–85; Smyth, "It Should be the Centre . . . .' "

83. To some extent, the difference may be due to the fact that many Ontario elementary schools include children up to age thirteen, while British primary schools stop at approximately age eleven. Given the historically based tendency for men to teach older children, there may be more men in the upper grades of the Ontario elementary schools, thus influencing the overall figure. See also note number 80.

84. Denys Giguère, "Gender Gap Widening among Ontario Teachers," *Professionally Speaking* (June, 1999): 42–45; Mahony and Hextall, *Reconstructing Teaching*. The trend is likely to accelerate as large proportions of older teachers will be retiring in the coming years and those are groups with higher proportions of men (Giguère). It is also tempting to conclude that the intensification of government control and scrutiny in recent decades means that those who have other choices (men?) might consider other careers a more attractive option.

85. Acker, *Realities*.

86. All names of teachers in both studies are pseudonyms.

87. Sharon Lapkin, "Introduction," in Sharon Lapkin (ed.), *French Second-Language Education in Canada: Empirical Studies* (Toronto: University of Toronto Press, 1998), pp. xix–xxx.

88. Ontario Ministry of Education, "French as a Second Language: Core French Grades 4–8," 1998. The document also specifies that the aim of the core French program is to give children a basic command of the French language, an understanding of the various linguistic structures in French, and a broader appreciation of French culture in Canada and in other parts of the world. Canada is an officially bilingual country (English and French), but in practice English is the main language spoken in all of the ten provinces and three territories except Quebec, although there are communities within other provinces that are mainly French speaking. Despite (and in some ways because of) the fact that core French is a required subject in Ontario, it holds a marginal position within most elementary schools.

89. Acker, *Realities*; Gertrude McPherson, *Small-Town Teacher* (Cambridge, MA: Harvard University Press, 1972).

90. Ontario English Catholic Teachers' Association [OECTA Survey: French as a Second Language], 1993, unpublished raw data.

91. Hargreaves, *Changing Teachers*, 1994.

92. Roland Barthes, *Mythologies* (Paris: Seuil, 1970).

93. Patricia Hill Collins, "Learning from the Outsider Within: The Sociological Significance of Black Feminist Thought," in Joan Hartman and Ellen Messer-Davidow (eds.) *(En)gendering Knowledge: Feminists in Academe* (Knoxville: The University of Tennessee Press,

1991), pp. 40–65; Sandra Harding, "Rethinking Standpoint Epistemology: What is 'Strong Objectivity?," in Linda Alcoff and Elizabeth Potter (eds.), *Feminist Epistemologies* (New York: Routledge, 1993), pp. 49–82.

94. Acker, *Realities*, p. 34.
95. Sandra Acker, "In/out/side: Positioning the Researcher in Feminist Qualitative Research," *Resources for Feminist Research*, 28, 1 and 2 (2000): 189–208.
96. Ralph Turner, "Sponsored and Contest Mobility and the School System," *American Sociological Review*, 25 (1960): 855–67.
97. Nias, Southworth, and Yeomans, *Staff Relationships*.
98. Biklen, *School Work*.
99. Hargreaves, *Changing Teachers*.
100. Marilyn Osborn, Elizabeth McNess, and Patricia Broadfoot with Andrew Pollard and Patricia Triggs, *What Teachers Do: Changing Policy and Practice in Primary Education* (London: Continuum, 2000).
101. Ibid., pp. 91–93.
102. Ibid., p. 97.
103. Hargreaves, *Changing Teachers*, p. 191.
104. Rita Kissen, "Forbidden to Care: Gay and Lesbian Teachers," in Deborah Eaker-Rich and Jane Van Galen (eds.), *Caring in an Unjust World* (Albany, NY: State University of New York Press, 1996), pp. 61–84.

CHAPTER 3

State, Gender, and Class in the Social Construction of Argentine Women Teachers

Graciela Morgade

SUMMARY

This chapter investigates the institutional and subjective processes of the feminization of teaching in Argentina. That is, the ways in which "the feminine" and "the masculine" of hegemonic sex-gender relations have intervened in the process of training Argentine women teachers and men teachers. At the same time, in a dialectic fashion, we will investigate how teachers, both women and men, have acted and/or act on themselves or on others (especially with respect to what I will later call the "school mother") by virtue of their unique appropriation of those relations. In summary, we will examine the processes through which women have become "schoolteachers" within the framework of sex-gender regulations that characterized Argentine society at the time the teaching field was socially constructed. The working hypothesis that will unfold in this chapter is that within the different environments where training occurs, teaching has tended—and still tends— to maintain the devices of feminization of modernity.

INTRODUCTION

The educational system in Argentina, as in the majority of Latin American countries, was born at the end of the nineteenth century. According to the beliefs of the political class of that era, the modernizing project of "building a civilized nation" in the vast national territory meant incorporating large masses of "rebel" natives and/or "uneducated" immigrants into schools that had to be created throughout the nation and which, until 1930, did not differentiate between the sexes. Other than offering home economics for girls and military training for boys, the curriculum was identical for both: what was most important was that children attend school.

From the outset, the work of teaching in the Argentine educational system was principally developed by women. In that era it was understood that teaching was a job "appropriate for her sex": women had already taught in the home and it would be "natural" for them to continue teaching in school.

The official call for women to enroll in the Normal School to be trained as teachers was a basic element of state policies that established the nation's educational system. In 1870 the first Normal School was formed (with secondary school-level courses to train primary school teachers) and enrollment was open to both women and men. However, it soon became clear that many men enrolled in its classes to gain access to better careers in the future, and only worked as teachers for a few years until they finished their studies and found a more socially prestigious and better paid job. As a result, in 1884 the state implemented a policy that focused on women. Normal Schools were created that were exclusively for women and scholarships were granted to girls from lower-class backgrounds who aspired to the higher social status promised by teaching. In each of the capitals of the provinces—the political division of Argentine territory—the urban center opened up new possibilities for girls. The doors to another type of secondary school education were closed to them— the "Colegios Nacionales," also created in the provincial capitals, which were intended to train the elite male ruling class that would continue its studies at the university.[1] Therefore, to a certain degree, the less socially prestigious Normal School had, by the fact of its creation amid contradictory liberal and conservative ideologies,[2] a certain democratizing stamp.

The call for girls to enroll in the Normal School found an extraordinary acceptance in civil society and in particular, among families. In a society divided into distinctly differentiated classes (traditional families from the colonial period and landowners and cattle ranchers on the one hand, peasants and urban working class on the other), some groups of immigrants began to develop small-scale businesses, industries or small enterprises, and increased their participation in politics. For these sectors of the population, the possibility of educating their daughters and improving their worth on the marriage market made the opening of the Normal School a stroke of good fortune. In this way, we can affirm that in Argentina, teaching was feminine "at birth," occurring earlier than in other countries where women slowly entered the teaching workforce.

Teaching became closely linked to the social condition of women in other ways. It also configured the lower-class "mothers" who came into contact with women teachers at school. The "school mother" was a subject who was equally defined by social relations in the school. And to a certain degree, the image constructed in the official pedagogical discourse about her characteristics intervened in the configuration of the teaching profession as a female job.

By analyzing documentary sources[3] and the life stories of retired teachers,[4] we will show how, for decades, schools tended to "feminize" mothers and women teachers based on the gender relations belonging to modernity—the hegemonic gender relations that were articulated with a world vision (also hegemonic) that belonged to the educated petit bourgeoisie.

The Social Production of Subjects in Schools

How does one "become" a woman teacher or a man teacher? How have women and men teachers arrived at what they are today? Far from a defined period with a beginning and an end, the production of a "woman teacher" or a "man teacher" in primary school education is a permanent process. The prevailing practices in the institutional cycle (training in the Normal Schools or in professional workshops) are supplemented by other training spaces that also intervene in the construction of the repertory of responses employed by teachers on a daily basis. An individual's own life experience with schooling and education is itself a formative space, as well as the "in service" training that occurs in schools.[5]

From this perspective, "becoming a woman teacher" begins much earlier than receiving a degree or obtaining a position.[6] "Teacher training" is produced through successive constructions, provisional and unstable, which combine elements contributed by other fields of individual and historical experience. In this complex process different social discourses intervene that are combined in what Foucault[7] has called the "technologies of power" and the "technologies of the self"; that is, social forms of domination and the ways in which we as subjects act on ourselves.[8]

We understand that gender relations are a constituent part of the technologies of power and of the self. In this respect, the hegemonic relations between "the feminine" and "the masculine" have intervened in the process of training a woman or man teacher in Argentina and, at the same time, in a dialectical form, these women and men teachers have acted and act on themselves and on others by virtue of their unique appropriation of those gender and class relations.

In her book *Nacimiento de la mujer burguesa* [The Birth of the Bourgeois Woman],[9] Julia Varela investigates the social processes that, in a convergent way, tend to make women "feminine" in modernity. Varela writes:

> By means of certain techniques and technologies of the government that were linked to the exercise of concrete powers and to the constitution of regimens of truth, the devices of feminization conferred specific qualities on the so-called feminine nature, and the devices of sexuality were articulated.[10]

and further on she adds:

> the expulsion of lower-class women from the regulated environment of the guild, the institutionalization of prostitution, the differentiated affiliation of women with legitimate knowledge and the expulsion of "bourgeois" women from Christian scholastic universities which provided access to the newly-born liberal professions, the strategic role played in the West by the institutionalization of the Christian marriage with its indissoluble nature, in short, the emergence of certain feminine styles of life, to which the humanists contributed in a special way by designing the utopia of the ideal Christian woman (the perfect wife), constitute indispensable pieces towards understanding the genesis of the devices of feminization.[11]

Varela takes up and specifies the concept of tributary devices from Foucauldian thought, a concept that refers to a particular combination of technologies of construction of social subjects. Foucault defines the technologies of self as matrixes of practical reason "which permit individuals to effect by their own means or with the help of others a certain number of operations on their own bodies and souls, thoughts, conduct, and way of being, so as to transform themselves in order to attain a certain state of happiness, purity, wisdom, perfection, or immortality." Technologies of the self are combined with others, in particular, with technologies of power, which "determine the conduct of individuals and submit them to certain ends of domination, an objectivizing of the subject."[12]

However, expanding on Varela's ideas, we can affirm that the formal educational system organized in the nation-state starting in the nineteenth century included "the devices of feminization" of modernity: in numerous countries in the Western world, among which Argentina is no exception, the construction of public education not only implied that a world vision would be imposed by one class sector but also, at the same time, that a certain gender order would be institutionalized.

On the one hand, the school represented a valorization of academic knowledge, male-centered and encyclopedic. The hegemony of scientific positivism was quite far from incorporating the knowledge traditionally created and communicated by women into the body of knowledge "worthy to be transmitted" through the apparatus of the school. This choice implied the systematic silencing of certain questions and as a result, their "private" nature was reinforced—in particular, sexuality. The systematic silences in school not only excluded "feminine questions," but also reinforced the assignment of these questions to a private world deemed opposite to the public world. In school, girls were the paradigm of "good behavior." The different ways in which girls were disciplined was also translated into different expectations about their performance: as long as a girl did not bother anyone, she was not a "problem" in class. Girls legitimized themselves in school by their personal effort, and a binary opposition became crystallized within the system: "intelligence" (masculine) versus "will" and "dedication" (feminine).

At the same time, to compensate for women's inclusion in the world of work, schools tended to exacerbate "the feminine" in women teachers and expropriate it from mothers. Women teachers accumulated maternal functions, leaving mothers with fewer of these functions. As we shall see, they turned into paradigms of ignorance, superstition, and negligence. Women were "better teachers" than men because they developed teaching skills "naturally," because they had no other legitimate space to perform paid work, and also, because their salary could be lower.[13]

However, and as added proof of the contradictory dynamic of a political liberalism that is closely linked to economic capitalism, the school, either by its actions or omissions, also critiqued the devices of feminization. Particularly in the bourgeois sectors, women's access to written language was, without a doubt, key to their inscription in the transmission and discussion of the knowledge of their times. The "public" character of school challenged to a certain degree the domestic project for girls and young women, which was the only possible outlet for broad sectors of the female population. At the same time, the "protected" nature of the school institution made it trustworthy for families who feared for the moral integrity of their daughters. The possibility of obtaining professional credentials also contributed to the promotion and diffusion of an egalitarian ideology and to the expression of silenced voices. In this last respect, the school was a space in which women teachers could carry out professional duties— a controlled, hierarchical space, but at the same time, a space in which she performed a remunerated job with social prestige.

Mothers and Women Teachers: Subjects Feminized by School

Once the Argentine educational system began to recruit women from upwardly mobile groups and lower-class families in 1894, the relations within the school among the adults having responsibility for the children generally featured women: women teachers or principals on the one hand, and mothers on the other. The abstract link between the community and the school has been, in the majority of its concrete manifestations, a relationship between women. Fathers have traditionally participated in the Cooperadoras Escolares [School Cooperatives],[14] but in a far greater way mothers have been—and are—the ones who congregate around the school door when children enter or leave, who bring the forgotten book or have "a word" with the teacher, who give a final kiss, or attend school events and meetings.

However, the process of construction and consolidation of the nation-state, and more particularly, the organization of the educational systems in the nineteenth century, implied, among many other transformations, an unusual interference by the state in family dynamics. The school competed with the arrangement of family power by virtue of the fact that it was an institution charged with transmitting certain cultural capital to young children. But education did not move in only one direction from the state to the family. Families also contributed their own stories and traditions, their unique ways of valorizing knowledge and understanding and applying norms, and their perspectives on gender relations. For this reason, the school has been a site of conflicts and negotiations between groups and individuals with respect to state, domestic, and parental power. It was one of the crucial factors in the close articulation between the state and civil society, or rather, in the process of "nation building" of modern states.

Following Pierre Bourdieu,[15] we understand the state as

> the *culmination of a process of concentration of different species of capital*: capital of physical force or instruments of coercion (army, police), economic capital, cultural or, rather, informational capital, symbolic capital. It is this concentration as such which constitutes the state as a holder of a sort of metacapital granting power over other species of capital and over their holders.[16]

And he continues:

> It follows that the construction of the state proceeds apace with the construction of a *field of power*, defined as the space of play within which the holders of capital (of different species) struggle in particular for power over the state.[17]

School has been—and is—precisely one of those social spaces where, despite struggles and contradictions, the state tended to "produce" its power, on the one hand, because it imposes a certain cultural capital, and on the other, by "absorbing" a portion of paternal authority by breaking its absolute supremacy.

The present-day naturalization of the obligatory nature of schooling is, without a doubt, a veil obscuring the severity of a measure that, since the end of the nineteenth century, required that parents delegate their authority over their sons and daughters on the intellectual, moral, and physical level. Universal education was placed above parental authority:[18] according to the Education Law No. 1420 passed in 1884, it was not only obligatory for all children to go to school between the ages of 6 to 14, but the state could call upon the police to guarantee a child's education.

In this way, the widening of modernity through the extension of citizenship toward "the other" (women, boys, and girls) in the domestic world implied the expansion of the "public" sphere. This world was broken up into individual subjects with established rights that were external to paternal masculine authority. Since, in this era, the "father" of the house was not only responsible for the children but also for all the women in the domestic nucleus, the universalist state discourse implied, among other things, a rupture in the patriarchal hegemonic order. However, this principle that was followed in many other kinds of state interventions,[19] had characteristics that were specific to education (and to public health), which distinguished it from the implementation of other rights.

When the right to vote was expanded, for example, a right was extended that only a few had previously enjoyed, but voting itself did not constitute a private practice before its expansion was passed into law. That is, the complete attainment of the right to elect and be elected by means of suffrage followed upon a practice that was already a public activity. However, the right to be educated (which also had a public dimension in the school), transferred a practice to the state that had already developed within the sphere of the family: the educational process already had a place in the home, even if the content was different.

The implementation of mechanisms that tended to ensure the exercise of this right affected the link between the state and fathers and mothers by virtue of the public/private nature of education. In its extra-domestic aspect, it challenged the father's power in the home, especially the concentration of his authority in "physical," material and bodily terms; fundamentally, his right to allow other family members to "leave the house" or not. In this respect, it was not the "educational" authority of the father that was at stake with obligatory education; instead, it was a certain national "trademark": the "poverty" of language, for example, or violent disciplinary practices, which were indirectly passed on to children by their fathers. On the other hand, the private dimension of "the right to an education" involved a specific and harsh gaze at mothers. The education of the family was in her hands.

Women teachers, therefore, not only had to teach reading, writing, and arithmetic, but also had to develop other skills more closely related to the domestic dynamic. It was up to them to convince fathers (and mothers) who were ignorant of the benefits of education, of its importance as well as their obligation to send their children to school. Once the children were in school, the teacher's mission was to instill the knowledge that was necessary for them to become citizens and workers. But to achieve that goal, she had to teach "civilized" habits—even cleaning and grooming the children herself when it was necessary—and transmitting the notions and feelings of a national identity to the immigrant or native masses.

A commencement address directed to the graduates of the Normal School in 1911 summed up the task that awaited the women teachers:

> It was necessary to arrive at school, overcoming obstacles that now seem insurmountable: to persuade the father of the importance of school so that he would decide to send his children; and then before teaching the child, to clean him and practically dress him; to carry out

duties in unhealthy wooden rooms, pretending that they were school-
rooms; to somehow multiply the few school supplies, to make draw-
ings, to be creative in every way, without forgetting, of course, the
other great task of attracting the masses to school, where the national
language was spoken, the names of Argentine heroes were pronounced
and the history of Argentina was told, which was unknown to the
majority of them.[20]

This mission, repeatedly proclaimed as "moralizing" from the time
schools were founded, not only signals a direction for teachers to take
with students; it implies work that includes their families as well. To
one of the founders of the Argentine educational system, Domingo F.
Sarmiento, *la moral* [morality]—which derives from "*moeurs*" or customs in French and from the Latin, "*mori*" (from which the Spanish
"*morada*" [dwelling] is derived)—is nothing but an abstraction based
on the habits of life followed in the home:

> Good manners in the well-to-do classes are a product of the home they
> live in, the grooming they practice, their sense of self-pride, the iron
> rod of criticism, a pleasant appearance, and ideas about morality and
> decency, which are common to all Christian societies.[21]

But these upper-class families had governesses at home. The
encounter between mother and teacher—apart from being mediated
by less social distance—took place on the territory of the former, who
moreover was the employer. The families that filled the schools came
from the lowest economic sectors—recent immigrants, and workers
with little skills and "education." Among administrators and, most
likely, among many women teachers as well, the belief seemed to exist
that the women-mothers of the students raised and educated their
children badly, insofar as health, morality, and habits were concerned,
and in their transmission of the ways of life and routine knowledge.

On the one hand, mothers seemed hardly dedicated to the education of young children: "They don't care." Or if they did care, they
didn't have time to show it. They worked all day in order to provide
the basics for those in their charge while living in crowded quarters
where the minimal necessities of daily life—water, fuel, food—were
scarce.[22] Formal education came along to rescue the children from
these desperate realities under the assumption that "mothers on the
whole are occupied full-time in caring for their brood, and ideal
mothers are scarce."[23]

The high infant mortality of that era, the result of minimal or non-
existent public health services, was one of the issues the schools could

indirectly "attack." In parliamentary debates on education in 1882, Congressman Varela presented his argument that women needed to be taught all subjects because they had to develop their maternal duties in a truly scientific way. For decades, his position represented the hegemonic mode of "thinking" about lower-class mothers:

> Is it men or women who are in charge of raising children to age five, years in which a third of all those born disappear? If the woman from the countryside knew about physiology and hygiene, she wouldn't throw herself at the mercy of every healer and quack. I'm sure that thousands of crimes can be counted, now and in the future, that are caused by the ignorance of mothers who, acting in the best faith in the world, have produced the horrifying statistics on infant mortality that I've spoken about.[24]

His perspective, widely shared by the different sectors of the government at that time, was that mothers have good intentions—of course they do, because they are good by nature—but are guided by superstitions not supported by academic knowledge. They reach out for help to whomever is nearest: grandmothers, healers, or close friends in the neighborhood who, with their "prejudices," do not prevent but rather increase digestive disorders, wide-scale contagions, and early deaths.

On the other hand, the pedagogical theory in ascendancy during the twentieth century opposed disciplinary methods based on physical violence and also called into question wide-spread practices of child discipline:

> Our brutish, bad, hypocritical, coarse students are the sad product of an education by brutish and bad parents, who don't care about the words they use in their children's presence. Maybe they think a child can be cured by increasing the beatings they receive and the vulgar language they hear at home? Remove from the system families that raise their children badly.[25]

Superstitious, dirty, violent, indifferent or overworked, the lower-class woman, it was believed, didn't "know," in terms of specific knowledge, the hows and whys of domestic work. The common knowledge-expert knowledge binary that characterized the creation of schools had its unique gender inflection with the introduction in 1912 of "home economics" for girls.

According to Marcela Nari,[26] this subject was added so that women could build a certain type of family structure and relationship that would turn family life for the men (husbands and sons), at the end of their work day, "into the preferred place for rest/leisure/relaxation."[27] For that to occur, thriftiness and good grooming would be the basic pillars of domestic organization, with proper ironing, an adequate kitchen, and a clean house and inhabitants.

Social class and gender are linked because the cosmovision and habits of the "upper classes" are not transmitted only by reading and writing (and the civil right to an education). Dazzling embroidery, delicate lacework, and magnificent tapestries were passed in front of the eyes of lower-class girls who worked long hours to learn how to be a good lady in society, rather than studying subjects relevant to their social reality and putting "backstitching, hemming, darning, dressmaking" to the side.

These signifiers and practices configured a form through which the school–family link was structured. The exaltation of "feminine nature" in the sector of women teachers not only tended to exacerbate their gender attributes but also, in a certain way, to alienate them from mothers. The feminization of teaching tended, paradoxically, to set itself in opposition to the "feminine" characteristics of the women-mothers, in a powerful symbolic operation that established hierarchies between women, and in this way, reinforced the power of the state as well.

GENDERED SUBJECTIVITIES IN THE TEACHING FIELD

However, during that prolonged initial period (which we can date from approximately 1870 to 1930), the feminization of primary schools in its quantitative and qualitative aspects did not meet any opposition from society and found very little open resistance in the practices of the majority of women teachers.[28] Even after 1930, when a military coup caused the first interruption to democracy, the educational system began to lose its supposed traits of "neutrality" in favor of certain "forms of ideological imposition, each time more obvious, dogmatic, and thereby more clearly authoritarian and coercive."[29] This advance in discipline—according to the same author—included, among other measures, an attempt to control access to teaching by women. But teaching did not lose its extremely high female participation (85 percent women in 1930, 84 percent in 1945), nor did it lose its relative importance vis-à-vis other alternatives open to girls in secondary education (such as secretarial studies).

The high rate of female participation was maintained in part by private intervention in teacher training (in the teacher training schools run by the Catholic Church) and in part by the determination of women themselves who were not willing to allow the most legitimate door open for secondary school studies to close. In some way, the tradition had already been firmly installed in families and in women. Unable to reduce the presence of females, the authorities had no choice but to intensify its control over women teachers.

THE CONSERVATIVE STATE IN THE THIRTIES

"the woman teacher": To Be, To Seem, To Obey

In the Thirties, to be a teacher was the natural destiny and first choice for women in socially and economically mobile sectors in Argentina. Guided by the advice of their parents, older brothers, other family members or local priests, or following social traditions or their own projects, these 13- or 14-year-old girls (whose parents had the economic means to send them) chose the Normal School for several reasons. They attended because it gave them a chance for mobility or social prestige, because of their desire for power or their love of children, or because they had a calling to serve the country or others. The widespread image of the "mother teacher" synthesized all of these aspirations.

A Normal School education was considered a complete and solid education. It went beyond the conceptual content that corresponded to different school disciplines: the school was a veritable academy for acquiring the habits of being and performing as a teacher. Clothes, ways of talking and conducting one's self in public, good working and organizational habits, along with other subjects that were more form than content, were an integral part of their education, as well as methods for teaching arithmetic or spelling. The construction of the woman teacher took place within a demanding climate that required hard work, often with an inflexibility and intolerance considered as necessary and beneficial. The discipline of the body and forms of behavior constituted a pedagogy in which one had "to be and to seem." "The woman teacher" had to set an example of morality, order, cleanliness, and work. In this sense, the ethos of the Normal School was absolutely clear:

> [In the Escuela Normal they taught us] not to make noise when we walked . . . during recess we weren't allowed to take books out because studying was done at home; you had to talk without raising your voice, you couldn't run, you were a young lady who had to set an example. (Juanita)

The power of the institution primarily rested on its academic prestige, since its male teachers had written books on pedagogy or literature, or were distinguished politicians or university professors. Its women teachers (who were few in number) were active, involved in education, and had occasionally written books themselves. Among them were also refined "ladies" who taught classes wearing gloves and consumed the "high culture" of the moment.

The Normal School foregrounded the girls' family and social status, leading them to deny their own background (if they were not from the upwardly mobile classes). A cosmovision was molded in which the "woman teacher" would play the protagonist's role in the construction of the nation and a modern society. And this cosmovision was reinforced by the everyday nature of the school. For many individuals, the school, along with the radio, was the only means of entry into the public world beyond the limits of the rural town.

Finding a job depended on luck and tenacity, but above all, on personal connections—knowing an administrator in an influential position, always male and obviously, from the governing party. It is the male who grants the opportunity to enter a system, which, in turn, is also controlled by males because, as in other countries, the base of the teaching pyramid is feminine but it becomes "masculinized" as it goes higher.[30] The absence of comprehensive work regulations, which was only achieved in 1958 with the passage of the Teacher's Statute that is still in effect, left women teachers defenseless against the prevailing politics of the day.

Their difficulty in finding a job corresponded to their low expectations when they accepted one. Jobs were generally found at a great distance from their family home, which meant teachers had to live in rooming houses or with other families. In contrast, however, they would not have been given permission to leave their families to study at a distant university. Their efforts to find a job earned them an unexpected social recognition: invitations to institutions that had previously been off-limits, better chances on the marriage market and, why not, a bit of pleasure:

> I was invited to a first communion and when I got there it was as if the Queen of England had arrived. All kinds of honors and other things that I see no longer exist today. . . . It was because the teacher was regarded as a supernatural being, and more so in those places where many people didn't know how to read or write. . . . Frequently they even said, "What an honor for me that you came." And I felt bad . . . it felt strange . . . but it wasn't really like that . . . what happened was that

in old times, the teacher was placed in another category. Also a teacher knew how to inspire respect. (Carlota)

But it was not always so friendly. Social reality collided with school routines and made each teacher's experience unique. In some areas, pregnant girls, dirty and lice-ridden children, or students who had jobs and slept in class, or others who were violent, presented a serious challenge to a teacher's work. The image of the woman teacher, however, had to be irreproachable: with a starched apron, stockings, and neat hair, teachers denoted with their impeccable body a distinctive social class and, in turn, a spotless morality.

Since women teachers obtained their first jobs at age 18 to 20, and often had students who were older than 14, the age set as the upper limit by the Education Law of 1884, the generational encounter was much more frequent than what we could imagine today:

> One day an inspector came and we talked in the patio. . . . I had never noticed that during recess the students were trying to make themselves look good for me. These were boys from 18 to 20 years old and I was 22. The inspector immediately realized what was happening and told me: "Miss! Don't you see the students combing their hair, fixing themselves up during recess? They're all in love with you. See how they look at you!" Things like that happened. While you were teaching they were like children, and you didn't realize they were men. (Maria Rosa)

The starched apron was not only meant to hide a sexed body. It also had to serve as a veil against external sexuality. Political conservatism also expressed itself in rigid moral precepts, and control over sexuality (after a certain freedom was allowed in the 1920s) constituted one of the core elements of the feminization of teaching.

No Set Working Hours and No Labor Rights

Teaching has traditionally been a badly compensated job. Nevertheless, during the 1930s salary reductions were made for the conservatives in power, a de facto tithe to the party that had given jobs to teachers. The effect of the salary reductions was mediated by the family configurations of the women teachers as well as their own low aspirations and scant labor consciousness.

Along with the erratic salary, few if any regulations existed that governed the working conditions of teachers. For example, classes tended to be large, with a minimum of 40 students. And if the children did not learn during school hours, it was considered normal for teachers to work after school. The teacher's house became an extension of the school: not only would she prepare classes for the following day, but she also provided free tutoring to students with difficulties.

The precariousness of the job, underscored by the uncertainties of the job search and the arbitrary way in which schools were run, was also evident during the exceptional moments in the life of a woman teacher:

> [The teacher was expecting her first child.] At that time you couldn't take maternity leave before giving birth. . . . we worked up to the last minute. One day I felt very bad. I never missed classes, always put up with things, but I asked the principal if she would let me go home. That woman, an embittered spinster who hated pregnant women, told me that before I could leave I had to help her in the library, and for a long time she made me lift and carry books. Finally she felt sorry because I looked so sick and she let me go. I arrived home and two hours later my son was born. . . . That's how the principal was. You couldn't miss classes. Unless she gave her permission, I couldn't leave. That's what it was like before. (Beatriz)

However, unlike women in other social groups, women teachers tended to keep working after their children were born or when these were young. The custom of living with their parents, the ease with which they could hire domestic help at a low wage, and the partial work day (in the morning or in the afternoon) made it easier to combine these responsibilities, and thus contributed toward the naturalization of the "mother teacher."

Stagnant Pedagogical Knowledge

The content and methods she learned in the Normal School and her own experiences as a primary and middle school student were, for the woman teacher, her most valuable tools when imparting classes.

Some, however, were troubled by pedagogical issues that arose out of the active school movement.[31] Despite the liberal roots of the movement, in Argentina, notably, the majority of those who adhered to its tenets were teachers sympathetic to the Communist Party and concerned about educational policies. However, the pedagogical movements of that era did not significantly undermine the signifiers of teaching nor its practices.

Women teachers cleverly thought of ways to somehow adapt their knowledge to their realities. On the whole they felt very confident of their objectives and the means to achieve them:

> Books and magazines arrived . . . *La Obra*, and the *Monitor de la Educación Común* was a magazine edited by the national Board of Education. Besides books, these guided us. We also received, from the National Postal Savings Bank, a very interesting book we used to teach the weekly class in banking. Stories, legends, poetry . . . . Each teacher had her book. The Monitor remained in the Library, as it usually had regulations. The principal would announce them at meetings. But it was so difficult to follow them! We followed what we could! We were in such remote schools, with students who were so different from those in the capital. I tell you, when I gave my first class, I didn't know how to show my authority in front of the class. (María Rosa)

To back up the work of the teacher, there was always the principal. As "the authority," it was understood that he or she would assume control of a conflictive situation. At the same time, his or her appraisals would earn the teacher "points," and the inspector would follow a teacher's career by reading these evaluations, which teachers could hardly contest.

In turn, the male or female principal was controlled by the male inspector, who revised the work of every teacher and gave a test to the students to see what they had learned, to be assessed according to a checklist. Thus, the male inspector (women did not rise above the lowest levels of the bureaucratic pyramid) controlled the principal, the woman teacher, and the children.

In this way, these subjugated subjects constructed a unique space in the history of work in Argentina. A knowledge centered on rigid paths, an arbitrary structure within which power was exercised, and a firm control over teachers' sexuality made teaching a space of social existence that was criticized only by a minority of politicized groups.

The Popular State

The arrival of Peronism in the mid-forties was a milestone in the lives of all women teachers. During Perón's government (1946–1955), which could be characterized as a populist movement in which the well-being of the people found its maximum expression, no substantive modifications of pedagogical methodologies were made; rather, there were changes in the ideological content of education, and fundamentally, in the relations at the base of the pyramid in the job sector.

The negative feelings against governmental interference in the curriculum were much stronger, for example, than they were against the salary reductions that were made in the midst of the conservative era. Women teachers felt and resisted the government's imposition in several ways:

> The changes were so great . . . . Because all the books, the principals, the meetings, everything revolved around Perón and Evita. Everything, all the laws, old age protection, books . . . . all the books. The students were required to read and discuss them. We always had to listen to the radio and read magazines in order to be able to teach the students. The principal controlled everything meticulously or else he would be sanctioned. There was no freedom. Yes or yes. If there was an official function, we had to gather the students in the courtyard, talk to them about the event and the reasons behind it. It was an order. I'm talking about our school. (María Rosa)

There was a growing sense that a teacher's work was being invaded, reaching a point of heightened tension in the mourning period after the death of Evita Perón:

> When the orders arrived, which couldn't be ignored . . . , teachers had to wear mourning for Evita. I read the orders to my students. I had some black ribbon, and I cut off a piece and put it on my apron. I told my students "you know that when my mother died I didn't wear mourning. Under these circumstances I have to put on this black ribbon. It's not a personal matter; there's an order requiring it." (María)

The relation between Peronism and the teaching sector was marked by conflict. A conflict that, in some ways, tended to pit the populist government against the educational sector in terms of class opposition: the expansion of primary school education reached a level it had never known before. However, from the teachers' perspective, the government extended education to areas that had been previously excluded only because it wanted to increase its power as an ideological machine. Once again the latent contradictions of liberalism put different social actors in firm opposition to each other. And not even the passage of the first Teacher's Statute (1949), which established regulations and acknowledged teachers' labor rights, was successful at building bridges between teachers and the government (with the exception of some Catholic teachers, for whom the introduction of religion into public education for a brief period of time served as a mitigating factor). Also, some Peronist teachers, generally daughters of working-class union members, experienced this moment with gratitude and intensity, becoming at times the party's "eyes and ears" to control their colleagues.

However, to interpret the passive and active resistance of women teachers as merely a form of conservatism within a class that viewed the expansion of the popular state with reluctance is to analyze the situation incompletely. During the fifties an increased female participation in the job market and the political sphere also began (women's right to vote was granted in 1947). Women teachers experimented with forms of opposition and began to question their supervision. In this context, the reinforcement of "the feminine" within schools, that is, the imposition of certain work relations and control, has a gendered component.

Thus, by the start of the fifties, a decade that was crucially important in the lives of women all over the world as well as in Argentina, the state and women teachers broke with the basic agreements that characterized work in education for at least six decades. Once this moment arrived, women teachers, on a massive scale, began to feel oppressed by official controls. In mostly hidden ways, such as seeking solidarity with students or, more infrequently, by joining unions or through open confrontation, they were part of a sector in classic opposition.

CONCLUSIONS . . . AND SOME BRIDGES TO THE PRESENT

Understanding the feminization of teaching as a dialectical process implies studying the technologies of power and of self which shaped women teachers, and at the same time, examining the ways in which those teachers contributed and contribute to the shaping of teaching as female work. The dynamics of gender and class that configured the relations of power that were intrinsic to the construction and consolidation of the teaching occupation in Argentina served as the basis for the characteristics that still persist in this sector: work in schools (especially at the classroom level) is performed by women who belong to the middle class (from the sociocultural point of view more than from the economic point of view).

Formal education tended to create "feminine" subjects among women teachers. At the same time, it provided them with a space for personal development with definite limits, yet their potential could not be entirely controlled. In a dialectical way, women teachers also tended to "feminize" teachers' work, and also to discuss it and mold it as they desired.

Since then, as in many countries in the Western world, a job in a school has signified both a space of autonomy and subjugation for women teachers.[32] A space of autonomy, because it allowed them to leave the paternal or marital home in order to perform a remunerated job and have a public presence; a space of subjugation, because the educational system, in its pyramidal and bureaucratic organization, kept them in classrooms and in schools, with few opportunities for professional development.

The mutually legitimizing relationship between civil society and the policies of the state which characterized the massive incorporation of women from the inception of the Argentine educational system began to show some cracks toward the middle of the twentieth century, caused to a great extent by the women teachers themselves. The gap between the definition the state proposed for teaching and how the teachers themselves defined their job, first noticeable during the Peronist governments, became more severe during the educational developments of the sixties,[33] when technocrats attempted to rationalize the job using principles from outside the teaching field. These changes, however, did not have a great impact on the hegemonic categories of gender and class in teaching.

At the beginning of the seventies, the "educational workers" movement[34] grew enormously, shifting the balance toward job consciousness. One result was a reduction in the feminizing content of the job. The end of the twentieth century found women and men teachers fighting for public education and against the growing limitation of social rights that neoliberalism appears to impose on a global scale. In this context, teachers discuss not only working conditions but also the relationship between knowledge and established power, thus interrogating the most powerful technologies of feminization of teaching. Nevertheless, the position of women within the same teachers' unions or the scarce thematization of gender relations on the part of the men teachers, indicates that the profession continues to struggle against the "mother teacher."

Becoming a woman teacher (or man teacher), insofar as it is a process in which subjectivities are constructed, can not be separated from the processes of childhood socialization, education in a specific institution, or the experience of working within a school organization (or in a union), except in an analytical way. Likewise, it is only in theoretical terms that it is possible to establish a distinction between the "feminization" of teaching and the "feminization" of women teachers as subjects.

NOTES

1. Andrea Alliaud, *Los maestros y su historia: los orígenes del magisterio argentine* (Buenos Aires: Centro Editor de América Latina 1993); Silvia Yannoulas, *Enseñar ¿una profesión de mujeres? La feminización del normalismo y la docencia 1870–1930* (Buenos Aires: Kapelusz, 1996); Graciela Morgade, *Mujeres en la educación. Género y docencia en la Argentina 1870–1930* (Buenos Aires: IICE-UBA/Miño y Dávila Eds., 1997).

2. While the liberals proclaimed the universalization of the rights of the bourgeois, the conservatives wanted to limit those rights to certain sectors of society.

3. Speeches and writings by Domingo Faustino Sarmiento and the parliamentary debates that surrounded the approval of the Education Law No. 1420 in 1884.

4. These are the life stories gathered from retired teachers aged eighty to ninety years old: Juanita, Carlota, María Rosa, Beatriz and María. I express my deepest gratitude to them for sharing their memories with me.

5. The bibliography on this topic is very wide. Among the works of my Argentine colleagues are Alejandra Birgin, *El trabajo de enseñar. Entre una vocación y el mercado: las nuevas reglas del juego* (Buenos Aires: Troquel 1999); Gabriela Diker and Flavio Zulma Terigi, *La formación de maestros y profesores: hoja de ruta* (Buenos Aires: Paidós, 1995); María Cristina Davini, *La formación docente en cuestión: política y sociología* (Buenos Aires: Paidós, 1995); Gloria Edelstein and Alicia Coria, *Imágenes e imaginación. Iniciación a la docencia* (Buenos Aires: Kapelusz, 1995).

6. Based on the reconstruction of the subjective processes of transformation that make an individual feel and conceive of himself or herself as a teacher, from recent investigations by Luiz Oliveira, "Trabahlo docente e subjetividade: embate teorico e novas perspectivas," *Revista da Faculdade de Educação*, São Paulo: USP, 12 (1998): 50–59. It is also possible to debate whether self-definition is necessarily produced at the moment professional accreditation is received.

7. Michel Foucault, *Tecnologías del yo* (Barcelona: Paidós, 1996, primera edición 1990).

8. From this perspective, the study of the formation of a teacher is increasingly related to studies on the construction of social subjectivities.

9. Julia Varela, *Nacimiento de la mujer burguesa* (Madrid: La Piqueta, 1997).

10. Ibid., p. 13.

11. Ibid., p. 14.

12. Michel Foucault, "Technologies of the Self," in Luther H. Martin, Hugh Gutman, and Patrick H. Hutton (eds.), *Technologies of the*

Self: A seminar with Michel Foucaut (Amherst: University of Massachusetts Press, 1988), p. 18.

13. Debates have existed over the amount that a woman teacher should earn, as it has always been regarded that what she earned must be inferior to what a man teacher should earn because it was taken for granted that she relied on her father or husband for economic support. In 1882, it was assumed that a man teacher should earn 800 pesos, while for a women teacher it was estimated that 500 or 600 would be adequate. (Graciela Morgade, *Mujeres en la educación* Myriam Feldfeber, "Génesis de las representaciones acerca del maestro en la Argentina (1870–1930)," Informe Final, Buenos Aires: CONICET, 1990).

14. Voluntary organizations that assisted in running the school building and cafeteria.

15. Pierre Bourdieu, "Espíritus de Estado," *Sociedad, Revista de Ciencias Sociales,* Buenos Aires: Facultad de Ciencias Sociales-UBA, 3 (1996), pp. 16–21.

16. Ibid., p. 10.

17. Ibid., p. 11. [English translations from Pierre Bourdieu, Practical Reason: On the Theory of Action, trans. Loïc Wacquant and Samar Farage (Stanford: Stanford University Press, 1998), pp. 41–42 Trans.]

18. In the area of health, vaccinations (although these were also controlled in school); in the area of law, the existence of legal defenders for minors and institutions to report family violence, and so on.

19. Meyer, Ramírez, Langton Walker, and O'Connor, in their now classic study on "The State and the Institutionalization of the Relations between Women and Children," in Sanford Dornbusch and Myra H. Strober (ed.), *Feminism, Children and the New Families,* (London; New York: The Guilford Press, 1988), pp. 137–158, nevertheless trace some variations of this general principle, in fundamental accordance with the dominant political and cultural system. In this way, they find three forms of modernization: in the more "liberal" countries (e.g. the United States) the visibility of family conflicts and thus the rate of state intervention is high; while those countries with a strong communitarian government (as in Scandanavian countries) generate intermediate levels of public discussion; and the systems which are denominated "organic corporations" (Latin American countries, for example, with a strong incidence of Catholic tradition and "natural" rights) tend to admit low levels of public perception of intra-familiar life and conflicts, and in this sense, the legitimacy of state action tends to be constructed more slowly.

20. Monitor de la Educación Común, Consejo Nacional de Educación. Buenos Aires, XI (1911): 543.

21. Domingo Faustino Sarmiento, in Héctor Recalde, *El primer congreso pedagógico* (Buenos Aires: Centro Editor de América Latina, 1987, p. 87).

22. Monitor de la Educación Común, Consejo Nacional de Educación. Buenos Aires, VI (1901): 905.

23. Ibid.

24. Monitor de la Educación Común, Consejo Nacional de Educación. Buenos Aires, I (1882): 538.

25. Monitor de la Educación Común, Consejo Nacional de Educación. Buenos Aires, IV (1888): 811.

26. Marcela Nari, "La educación de la mujer," *Revista Mora del Instituto Interdisciplinario de Estudios de Género*, Buenos Aires: Facultad de Filosofía y Letras-UBA, 1 (1995): 35–41.

27. Ibid., 25.

28. Graciela Morgade, *Mujeres en la educación*, 1997. In this publication I have analyzed documentary sources from the period in which hegemonic groups such as those critical of the liberal economic and political system of that time, refer quite noticeably to the social role of women as "mother teachers," characterized by their capacity to "create" life, socialize children, and care for "others" in the domestic atmosphere while at the same being subordinate on the intellectual and political plane.

29. Juan Carlos Tedesco, *Educación y sociedad en la Argentina, 1880–1900* (Buenos Aires: Ed. Solar, 1988).

30. Charol Shakeshaft, *Women in Educational Administration* (Newbury Park, CA.: Sage Publications, 1991); Kerreen Reiger, "The Gender Dynamics of Organizations: A Historical Account," in Blackmore, Jill and Jane Kenway (eds.), *Gender Matters in Educational Administration and Policy: A Feminist Introduction* (London; Washington, DC: Falmer Press, 1993); Jill Blackmore, *Troubling Women: Feminism, Leadership, and Educational Change* (Buckingham; Philadelphia, PA: Open University Press, 1999); among others.

31. Based on the works of Montessori or Dewey.

32. Madeleine Grumet, *Bitter Milk: Women and Teaching* (Amherst: University of Massachusetts Press: 1988); Sandra Acker, *Teachers, Gender, and Careers* (New York: Falmer Press, 1989) and *Gendered Education: Sociological Reflections on Women, Teaching, and Education* (Buckingham; Philadelphia, PA: Open University Press, 1994); Alison Prentice and Marjorie R. Theobald, (eds.), *Women Who Taught: Perspectives on the History of Women and Teaching* (Toronto: University of Toronto Press 1991); Michael Apple, *Maestros y texto. Una economía política de las relaciones de clase y sexo en educación* (Barcelona: Paidós, 1995).

33. In that period, official educational proposals advocated the professionalization of work specifications: the consumer–product paradigm

was the model for thinking about the relationship between education and society (annual, monthly, and daily planning with objectives, the psychologizing of relationships with students, etc.).

34. This movement began in 1973 with the founding of the Confederation of Educational Workers in Argentina (CTERA). This organization affirms that teaching is a job, not a ministry, and teachers should fight for their rights as workers.

PART II

Multidisciplinary Interpretation of the Feminization of Teaching

CHAPTER 4

Women Teachers in Mexico: Asymmetries of Power in Public Education

Regina Cortina

When Mexico instituted public education in the early days of the twentieth century, the teaching profession was a male-dominated occupation in most of Latin America. Men quickly abandoned the classroom, however, upon finding new opportunities for political and labor organization in the state bureaucracies taking shape in those same years.[1] As a result, throughout the twentieth century, especially when the expansion of urban and rural schools was at its peak, thousands of women entered education, thus bringing about the feminization of teaching and the emergence of power asymmetries between men and women in public education. At the present time, education in urban schools is still an occupation with a disproportionately high female presence in Mexico as well as in other Latin American countries. Even though there is equal pay for equal work, the differences in the average salaries and job responsibilities of men and women in the educational system continue to be a reality.

Most women workers in Latin America today are employed in occupations dominated by women. These occupations are characterized by the lack of professional autonomy and a low salary.[2] Over the past few decades, teaching has become the female profession *par excellence*, with a low salary level and a significant presence of men in administrative and leadership positions. Moreover, the high rate of feminization in the profession[3] has had an influence on the self-esteem of teachers, and it contributes to their low professional status.

In the majority of nations, teachers can be found at the base of the educational pyramid.[4] Teaching has been considered women's work for decades, with an organizational structure differentiated by gender and a low level of professional autonomy. The feminization of teaching affects both women and men teachers. The lack of autonomy among women teachers is clearly reflected in the fact that their work is supervised by school principals and higher authorities, such as school supervisors, public bureaucrats, and ministers of education, most of whom are men, who determine, among other things, the curricular content that must be taught for each subject. The asymmetries of power that are present today in teaching reflect the different social and cultural value given by society to men and women over time.

In the contemporary history of Mexico and Latin America, a feature shared by many women who distinguished themselves in public life was having been a teacher. In the first decades of the twentieth century, especially in the 1920s and 1930s, years in which nationalist movements triumphed in countries such as Brazil, Mexico, and Argentina, the leaders of these movements and their ideals were focused on masculine aspirations to consolidate military power and design authoritative national governments. Yet it was women who worked to achieve social reforms and expand civilizing ideals to rural areas. Lacking in sufficient numbers in domains where power was being concentrated and redefined, they were not able to project their contributions into the public discourse of nation-building.

During the period when governments recruited more women for teaching positions and increased the number of schools, support networks were created among women teachers that allowed them to expand their ranks in the only profession that had been open to women since the nineteenth century.[5] Women teachers participated actively in public life, making contributions as journalists, early feminists, union members and poets. Most notable among them were Gabriela Mistral, Rosario Castellanos, María Lavalle Urbina, and all the women teachers who mobilized themselves and played an active role in the First Feminist Congress of Yucatán.[6]

At the beginning of the twentieth century, women teachers belonged to the earliest group of middle-class educated women in Latin America, particularly in Mexico. They were the first to challenge, in their newspaper articles, essays and books, the economic, social, and legal inequalities faced by women on the job market.[7] By studying the woman teacher, one gains a broader understanding of the cultural, social, economic, and political life of the nation. Existing studies, which are written using universal masculine pronouns, help erase from the historical memory the energy and political militancy among women who sought to achieve social reforms and increase their public presence.

Between 1880 and 1940, a feminist movement developed in Mexico that incorporated thousands of women from all social classes. "Most Mexican feminists came from the middle class and had to work for a living," writes Ana Macías. "Most of them were elementary school teachers. . . ."[8] Mexican women teachers had a prominent role at the first Feminist Congress of Yucatán in 1916, an event at which, for the first time, Mexican women came together to discuss their social and political rights, their right to an education, and the need for an educated citizenry.[9]

A key moment in the political life and public participation of women teachers occurred during the 1930s when women organized to form the *Frente Único ProDerechos de la Mujer* [United Front for Women's Rights] to obtain citizenship and the right to vote,[10] which would not be secured until 1953. Even by the 1930s, though, women teachers had achieved greater job security because they were state employees. The teaching profession gave women a stable job with equal pay for equal work if they were in the same professional category as men, unlike other occupations where women earned less than men who had the same level of education. Salary differentials, however, continued to exist because of the high concentration of women in classroom teaching and their low representation in administrative and leadership posts.

Despite their political mobilizations in the past, a social identity established at the beginning of the twentieth century strongly adheres to the professional and union participation of women teachers today. This social identity is endemic to the teaching culture and has hardened the gender and power asymmetries that determine the political and professional participation of women teachers.

The following pages will explain how a professional archetype of the woman teacher took shape through the political discourse that created the national educational system in Mexico, and show how the gender ideology that assigned gender-specific spaces to men and women reinforced a hierarchy within the profession and the national teachers' union. By recovering the public presence of women through writings and political speeches, this chapter explores how the social identity of women teachers was shaped and then used to define their work as a service vocation.

Interpreting historical documents and academic studies of occupational segregation by gender among women and men teachers, I will first describe how the ideal of the woman teacher was constructed. I will suggest that the attitudes and values toward women that were evident in the expansion of schooling need to be reconsidered in order to professionalize the pedagogical and professional function that women perform. Second, I will describe initiatives focused on the creation of a new teaching culture that aim to project a different image of women teachers in public education. Third, I will analyze three strategies designed to solidify the presence and professional participation of women teachers through educational policy-making.

THE PUBLIC IDENTITY OF THE WOMAN TEACHER

In the years following the end of the Mexican Revolution of 1910, when the Mexican State was consolidated and public education expanded, José Vasconcelos, the first Minster of Education of the post-revolutionary government, capitalized on the visit of Gabriela Mistral to promote the entrance of women into teaching as a dignified and useful profession. Women's participation in a salaried job was thus legitimized. The legacy of Mistral and her vision of teaching as a service vocation and a female occupation has continued until the present, unmodified by time, and is reflected in women teachers' self-perception as servants of the State, rather than as professionals who are in charge of their own teaching practice. The fiscal crisis of the 1980s in Mexico and the economic measures taken in its wake had a devastating effect on the teaching profession, causing it to lose its status as a middle-class occupation. As a result, women and men in teaching, who are praised at a rhetorical level, encounter little financial support to strengthen their pedagogical role within the classroom and improve opportunities for professional training, continuing development, and higher salaries.

Gabriela Mistral,[11] also known as "The Teacher of the Americas," helped shape the national identity of the woman teacher, who at that moment in history had become the principal actor in the service of the Mexican state, a representative of the legitimate national culture that would be consolidated through the school, and a model of citizenship. Gabriela Mistral occupies a central place in the history of education in Mexico. A poet and educator, the first Latin American woman to receive the Nobel Prize,[12] Mistral arrived in Mexico a year after José Vasconcelos created the Ministry of Public Education. As Minister, he directed the federal program to unify the nation by expanding public education, a project of enormous importance in the consolidation of public institutions and political life in modern Mexico.

Mistral's visit to Mexico in 1922, along with her writings, contributed to the creation of the social archetype of the "woman teacher" in Mexican society. Her stay coincided with the massive and permanent influx of women teachers into the Mexican public educational system. José Vasconcelos asked Gabriela Mistral to come from her native Chile and help him promote the educational reforms he was introducing. Those reforms sought the establishment of schools; the expansion of cultural missions in remote rural areas which preceded the creation of the Mexican rural school; the opening of libraries; the public recognition of the muralist movement as an expression of cultural nationalism; and the creation of a cultural elite.

Gabriela Mistral was invited to Mexico in order to be honored as a poet and educator. Vasconcelos capitalized on her visit to inaugurate the *Gabriela Mistral* vocational school for women, where students were taught practical skills such as typewriting, sewing, cooking, and pastry making,[13] in addition to the basics of language and literature. The objective was to give them technical skills so that they could discover their vocation and would not have to "follow a career for which they had no inclination."[14] Women were offered the chance to be educated in the so-called female occupations, designed as such because they would complement the roles of women as mothers and housewives. The foundation of this school, along with the publication in 1923 of Mistral's *Lecturas para Mujeres* [Readings for Women],[15] were events intended to promote the entrance of women into teaching. These developments came just as the Mexican State needed a great many people to spread the ideals of educational reform.

The innovations championed by Vasconcelos represented a step forward in the secularization of education, which had been initiated under Benito Juárez's liberal government (1858–1876). Under the liberals, elementary schools were opened for both sexes and secondary schools were established with different curricula for males and females. Secondary school was the highest educational level for young women, unlike preparatory school, which made it possible for young men to continue their studies at the university level. Such were the limits placed by politicians and educators on the professional aspirations of women.[16] The conception of women held by the liberals and their ideals for education and socialization—to compel girls and boys to accept pre-established models of femininity and masculinity—were perpetuated in post-revolutionary Mexico through the idea of the "woman teacher," treating teaching as a female vocation that complemented the domestic role of women.[17]

In 1921 the Mexican Revolution ended, followed by the most significant decades in the construction of the contemporary Mexican state, with important cultural and educational developments and the strengthening of nationalism. Because of the geographic diversity and the great plurality of the rural and indigenous populations—with the latter often unfamiliar with Spanish and speaking a wide range of indigenous languages—it was believed that social cohesion and the development of a national identity could be achieved by means of literacy and instruction in the Spanish language.[18] Through public education, Vasconcelos sought to use the Spanish language to implant a cultural model that was in keeping with the consolidation of political power.

Vasconcelos, who like Mistral was a humanist, was convinced that a national culture could be built through education, and a literacy campaign was begun to eradicate illiteracy. Along with an increased number of public schools, he believed that the wider availability of books would help educate the great majority of Mexicans. To this end, he created public libraries, known as the Bibliotecas Populares,[19] so that Mexicans could educate themselves. In each of the nearly one thousand libraries opened in that era (1922–1929), Vasconcelos included books by Mexican and Spanish authors, as well as classical philosophers and writers such as Aristotle, Descartes, Kant, Shakespeare, Rousseau, and Voltaire. The editions in Spanish of these classics are still considered to be among the most beautiful publications that exist of these books, which are commonly referred to as "Vasconcelos's green books."

Vasconcelos's high regard for public institutions was not shared by Mistral, who had little respect for authority and doubted whether learning could take place in educational institutions, perhaps as a result of her own autodidactic experience. For Mistral, education meant transmitting values and absorbing the classics so as to avoid perpetuating institutions. Thus in *Magisterio* and *Niño* she affirms that "School must make people hungry for culture; it's worthless to bestow it on them."[20] In addition to her lack of faith in institutions, Mistral's vision of teaching as a calling, and not as a profession, undoubtedly inspired a massive influx of women to preschool and elementary education. Although her stay in Mexico was brief, her presence and thoughts lived on in SEP (Secretaría de Educación Pública) publications that were intended for teachers, such as *El Maestro Mexicano*,[21] where a section called "The Mexican Woman Teacher's Page" praised housework as an important part of a woman teacher's duties.

In Mexico, "where no woman is as beloved as you," as the Minister of Education wrote to Mistral,[22] the Chilean teacher fulfilled an essential function during the implementation of Vasconcelos's policies. Mistral helped Vasconcelos articulate his policy of *mestizaje*, the mixing of races, and she helped to shape the Mexican teaching profession in her writings when she created the symbolic image of the woman teacher as an agent of the State, whose goals were to advance social justice by blending the races, creating a national identity, and forming a new political structure.

Gabriela Mistral, like many other Latin American women from poor backgrounds, began her intellectual career as an uncertified rural teacher, since the Normal Schools were concentrated in urban areas and were mainly for middle-class women. Later she received a degree and worked as a secondary school teacher, ultimately becoming principal of a *lycée*.[23] In her works she identified strongly with her teaching vocation and described pedagogy as a feminine art.[24]

Mistral believed that the interests of men and women were quite distinct. She advocated a differentiation of masculine and feminine occupations and interests. For her, "the child directs a woman's duties like a guiding light,"[25] and for this reason she commented on the lack of textbooks in basic and vocational education for women. The aim of women's education, as expressed in the document that created the *Gabriela Mistral* School, was "to equip [women] so that they can acquire a greater aptitude for the work, labor or duties they perform, provided that they always learn by doing."[26] The curriculum it followed derived from the characteristics of schools during the liberal governments, exhibited through its emphasis on values and not on knowledge. The goal was to create, through the school, a social ethic of national unity, and to form the values and aptitudes that would guarantee the social, cultural, and economic development of the nation.

During that era in Mexico, there was a great controversy among teachers over the fact that a "foreign teacher" had been invited to represent the "Teacher" in Mexico, and that a school had been built in her honor.[27] Sensitive to the criticism of her Mexican colleagues, Mistral decided to shorten her stay. Before leaving in 1924, she wrote "Words of a Foreigner," an introduction to her collection *Lecturas para Mujeres*, which she dedicated to the students of the *Gabriela Mistral Vocational School*. Mistral wrote, "This is the first attempt at a project I will one day realize in my country, destined for the women of the Americas. I feel they are my spiritual family; I write for them, perhaps with little preparation, but with much love."[28]

For Gabriela Mistral, her visit to Mexico had a great impact . It increased her visibility as a commentator on social and political problems in Latin America. Above all, it deepened her passion and commitment to indigenous communities and social justice.[29] So it was that she became known as the "Woman Teacher of the Americas" and one of the first Latin American women with an international reputation as a public intellectual in a world dominated by men.

"The Woman Teacher" in the Mistral's Thoughts

"Pedagogical work is a vertical vocation," wrote Gabriela Mistral, "not simply a matter of a position and a salary."[30] For her, teaching was a service vocation, a spiritual calling that to a certain degree had religious overtones. In his book, *La vocación vertical*,[31] Álvaro Valenzuela performs a detailed analysis of various texts by Gabriela Mistral concerning her vision of the teacher and the use she makes of the term "vertical," which according to Valenzuela is always associated with something positive and spiritual. It is important to add that at no moment did Mistral regard teaching as a professional career or as a remunerated occupation for women. A woman's presence in the labor market could be justified because she was holding a job related to the care of children and compatible with domestic roles. Once the State could count on the availability of women, it was able to meet the challenge of universal schooling.

The image of the woman teacher that Mistral created is essentially maternal,[32] influenced by the archetype of the mother and filial love: "If you can't love, don't teach children."[33] In *Magisterio y Niño* [Teaching and the Child], Mistral expanded on this relationship: "the teacher is a secondary mother, sometimes a corrected and enlarged one. This hybrid, a doubled being composed of knowledge and love, is a pure marvel, a kind of archetype."[34] Influenced by Froebel, she saw the teaching profession as a supplement to the ideal of motherhood.

Regarding the social and public impact of the teachers' work, Mistral dedicated many pages to exploring the problems of education. In them she repeatedly emphasized that the future of the Americas was in the hands of teachers, declaring: "the teacher must be the priest of the new religion that worships the *patria*, with the school its temple and the book its ritual."[35] Concerning educational reforms in Mexico, she wrote with great enthusiasm about Vasconcelos's commitment to promoting change in post-revolutionary Mexico, and she supported his efforts to use education to incorporate indigenous groups and cultures into the nation.

At the historical moment when the idea of a nation was being consolidated and the foundations were being laid for the rapid expansion of public education, Mistral's contribution was to equate teaching with a service vocation, one that was founded on filial love and the image of the mother, in which the duty of the individual had great social influence. This idea accompanied and supported the establishment of gender and power hierarchies within schools in the early decades of the twentieth century. These hierarchies, which still persist today, have limited women to the classroom, only allowing them a reduced political and professional role outside it, while men exercise political and administrative leadership.

The teaching profession in Mexico evolved with close links to the State and to the educational reforms that originated in the different political regimes that followed the Revolution. Starting in the 1940s, this relationship was formalized with the creation of the National Union of Educational Workers (*Sindicato Nacional de Trabajadores de la Educación* or SNTE), an organization closely tied to the political control of the Institutional Revolutionary Party (*Partido Revolucionario Institucional* or PRI). More than a third of all public sector employees in Mexico have joined the SNTE; its present membership is approximately 940,000, of whom roughly two-thirds are women, a percentage that is much higher in urban areas where women constitute 80 percent of the rank and file membership.[36]

From the Mother-Teacher to State Worker

Founded in 1943, SNTE is the union with the highest membership among all the unions for state or federal workers; it is also the biggest union in Latin America. Within the union different groups exist that are closely linked to the power of the PRI. In my book *Líderes y construcción de poder: Las maestras y el SNTE*, I show how one of the groups in power, *Vanguardia Revolucionaria*, accorded a specific place in the union culture to women. I give the name "social syndicalism" to the tactic developed by *Vanguardia Revolucionaria* to restrict women and women teachers to what are essentially social activities, specifically, welcoming them to union headquarters for breakfasts or on Mother's Day, inviting them to educational conferences, or allowing them to organize book fairs. All these strategies limit the participation of women in union leadership and political life.[37]

In 1989 when the first woman assumed a powerful position within the union, nothing changed for the thousands of women who were its core membership. Elba Esther Gordillo, known as "*La Maestra*,"[38] was the first woman named to lead the SNTE as Secretary General, but she has been able to remain in power although her term of office officially ended. She is now President[39] of the union and the Secretary General of PRI. When the writers Sabina Berman and Denise Maerker went to union headquarters to interview Elba Esther Gordillo for the television show "Women and Power," they observed:

> Men, men, men: from the moment we entered *La Maestra*'s building until we left, hours later, with the exception of Gordillo, there was nothing to be seen but men.[40]

The SNTE is a pyramidal structure, in which an increase in power means a decrease in women's participation. For the majority of women who form the base of the union and the teaching profession, opportunities for leadership and political representation are still not present. The hierarchies of power within the teaching culture have hardly changed since Mistral's visit in the 1920s. Although Elba Esther Gordillo is the first female leader of a union whose base is dominated by women, her leadership within the SNTE has been characterized by her promotions of a majority of men to positions of power. As a result, women teachers have excluded themselves by not pursuing positions they view as masculine. Although she has supported the magazine "*La maestra*,"[41] Gordillo has only paid lip service to women's greater participation in the union. The incorporation of women into her political discourse is rhetorical and has not substantially benefited women teachers either economically or professionally.

With Gordillo's ascent to power, the rhetoric of the union has centered on gender equality. A comprehensive program to address the needs of the woman-teacher has been created. Among its more notable achievements are that women no longer have to prove they are not pregnant when they join public service; and ISSSTE[42] medical services are provided to their families, including their husbands, who were previously excluded from coverage (in contrast, the wives of men teachers have always been covered). But much remains to be done, such as providing access to day care for the children of all women workers, opening up more leadership positions to women and expanding their political participation—a difficult goal to achieve as long as women provide the domestic support for men as they ascend the career ladder.

From State Worker to Professional Woman

In Latin America, and particularly in Mexico, efforts have been focused on changing the teaching culture and the social identity of the woman teacher. Within the framework of the agreements of the World Conference on Women in Beijing in 1995[43] and the Convention on the Elimination of All Forms of Discrimination Against Women,[44] there has been a growing awareness that the legacy of traditional socialization and its effect on women's social and political participation are the principal obstacle. International treaties place a special emphasis on education, as educational systems are crucially important to creating new perspectives on the role of women and men in society. In this way, attempts can be made to change prevalent attitudes and expectations about girls' and boys' education with respect to their cognitive skills and abilities, areas of academic concentration, potential for leadership, and future contributions to society.

Efforts to incorporate gender and gender equity into teacher training and the curriculum are mainly generated by nongovernmental organizations (NGOs) financed by international organizations and foundations.[45] Although examples are few, they represent attempts by NGOs associated with the women's movement to influence formal education. An area of crucial importance is to improve awareness among teachers of their function in promoting equality of opportunities for children through education.

Studies done on this subject in Latin America during the last decade show that teachers themselves greatly resist reflecting on these asymmetries and the role they play in their reproduction. For example, in a study done in Colombia, school personnel attended a workshop in which they were asked to reflect on the forms of discrimination against women in educational institutions. The authors reported that in many cases, the administrators not only lacked theoretical knowledge, but were also unaware of the concepts used to describe discrimination against women. In one instance, many participants responded that "sexism" was "a term they hadn't heard of before, nor could they associate it with anything." A small number of participants, the majority of whom were men, "linked the concept of *sexism* with *discrimination*."[46] In their conclusion, the authors indicated that as a first step to identifying the causes of sexism, school personnel needed to familiarize themselves with the concept of sexism and broaden their vocabulary in order to name sexist practices within the school environment. As the authors pointed out, the resistance of school personnel to this topic is very strong.[47]

In Paraguay, a project to familiarize school personnel with the concept of "gender" produced resistance and difficulties because of staff perceptions that such a discussion would undermine the traditional function of women and question the socializing function that attributes to mothers and women teachers (second mothers) an indispensable role in the socialization of children and the formation of the family.[48]

In the case of Mexico, there are several examples of collaborations between the State, universities and NGOs. A series of projects focus on providing spaces for women teachers to reflect on their work and on eliminating sexist images and language from textbooks. These are all attempts to modify traditional socialization through the school to favor equality between the sexes, so that women will be able to compete in the job market on a fairer basis. Evaluations of these projects have yet to be done, but the results of any such evaluation should be taken into account when preparing new manuals and courses for teachers about the gender components of classroom dynamics, and the relationship between the school and the family.

The importance of these projects over the long run will be in their capacity to influence the teaching culture that currently exists and to introduce into the school curriculum new perspectives on the equality of opportunities for men and women, given that in the majority of formal educational systems in Latin America, starting with preschool, different educational abilities are emphasized for girls and boys. Gender stereotypes are reproduced in both formal educational systems as well as in families, resulting in a high concentration of women in professions such as teaching and nursing.

At the university level, the Gender Studies Program at the Universidad Pedagógica Nacional de México (UPN), has introduced a seminar in which women and men teachers can reflect on and analyze aspects of gender equality as it relates to their own teaching practice. A total of 105 women and men teachers have taken these courses in professional training in the state campuses of the Universidad Pedagógica Nacional.[49]

Although official guidelines regarding curricular modifications do not exist in many countries, it is worthwhile to review the pioneering advances that have been achieved. Among the organizations and projects in this area is a program in Mexico City called "*Otra forma de ser maestras, madres y padres*," which gives women teachers the opportunity to work with mothers and fathers in an alternative educational project aimed at finding other ways to educate and take care of children in the preschool system. The project was first conceived as a voluntary collaboration between twenty-five women teachers, principals, and supervisors in several preschools in Mexico City.[50] In its subsequent phase, the Ministry of Public Education provided financial support for classes and seminars for four hundred teachers, principals, and supervisors. The next phase will be its institutionalization in teacher training schools.

The inclusion of a gender component in education depends on the interest of women and men in their own professional development. A research study in the state of Puebla provides us with a detailed profile of the teachers in that state, including a description of their social and economic situation and working conditions.[51] This study analyzes the differences in the level of initial training and the socioeconomic status of teachers who work in middle-class urban zones, marginal zones, and indigenous zones. The majority of teachers in the state of Puebla are women, and those who possess a higher level of education work in middle-class urban zones. However, two interesting facts need to be underscored: most of these middle-class women teachers indicated that they chose their career in order to follow a "service vocation"[52] and they also mentioned that they had not taken any classes in professional development in the previous five years.[53] These two statements confirm that the image still endures of teaching as a vocation especially suitable for women, and that it is also seen as a service vocation that does not require a constant updating of professional skills. In contrast to the concentration of women in urban zones, we find a higher concentration of men in rural zones and in school districts that serve indigenous groups.

From Rhetorical Recognition toward a Professionalization of the Teaching Function

One of the weaknesses of post-revolutionary educational development in Mexico has been the lack of genuine support for the teaching profession. The impoverishment of teachers beginning in the 1980s is the result of structural adjustment to economic and fiscal policies that had failed and the government's decision to reduce spending by cutting wages rather than reducing employment among civil servants. Teachers bore the brunt of these measures, as their salaries remained lower than those of all other state employees. In 1995 the average salary for teachers was less than half, in constant dollars, of what it had been in 1981.[54] In 1988, the worst year of the crisis, salaries for teachers fell 78 percent, while salaries for other state workers fell 55 percent.[55] These statistics show that women teachers in Mexico (who account for two-thirds of all teachers in the country) were the ones who suffered the greatest cutbacks during the crisis, while the union leadership and its president, Elsa Esther Gordillo, benefited under authoritarian PRI politics. At the same time, the federal budget for education in 1988 sank to only 22 percent of its level in 1981; it was only in 1993 that it reached the level it had had in 1982 in real terms.[56]

Structural adjustment policies that started in the 1980s have had a devastating effect on the teaching profession. Young people are not interested in teaching and the average age of teachers in the country is rising. In a study done in Mexico City, it was found that only 1 percent of teachers were under age 24; 25 percent were between ages 25 to 39; and 40 percent were over age 40.[57] This problem is not exclusive to Mexico. One of the main concerns in many countries in the world is how to enroll better-educated candidates in teacher education programs. The lack of interest among young adults in becoming teachers affects the viability and the future of the profession, and will have a strongly negative effect on the competitiveness of the Mexican economy in global markets.

The fact that the teaching profession has lost its status as a middle-class occupation in urban areas is related to the inadequate education of most teachers, who work without the necessary training and without a university degree. Of those teachers working in public education in Mexico City, only 3.5 percent have a university degree and the rest have the equivalent of a *bachillerato* (the Normal School program of three or four years).[58] The inferior academic background of the women and men at the front of the classroom has a detrimental effect on schools attended by children of poor families, but the problem does not stop there. The low academic level of teachers affects all sectors of society.

Moreover, the work of teachers is in reality a part-time job. Teacher employment in Mexico is characterized by a 25-hour per week schedule (unlike most countries where teachers work 40 hours per week), very secure employment, and a powerful national union that defends job security and stability. In Mexico teachers are paid only for their work during class time. Time spent in class preparation or in teacher development is not considered part of their job. Furthermore, Mexican teachers work two shifts, which means that the average number of students they teach is much higher than what is pedagogically advisable, especially in high schools. The practice of the "double shift" that was introduced by the union during the 1980s to obtain power and loyalty among teachers has become a significant obstacle to the professionalization of teachers. In Mexico City, 50 percent of teachers work two shifts and 60 percent have two occupations.[59]

In his analysis of the traditional lenses through which teachers are viewed in Latin America, Juan Carlos Tedesco has identified three categories: a merely rhetorical recognition of the importance of teachers; teachers as the victim or the party responsible for the poor functioning of the system; and finally, the devaluation of teachers, and their replacement by books or technology. Unfortunately, in Mexico teachers have been traditionally given a "merely rhetorical recognition of their importance"[60] by educational policy-makers, who do not back up their words with programs to help strengthen the function and importance of teachers in the educational process.

Current educational research shows that the work of teachers can be strengthened through public policy by means of three interrelated dimensions: opportunities for initial training and ongoing professional development; work conditions; and the salary and benefits received.

The initial professional education for most teachers still takes place in Normal Schools, which over time have become a second-class educational system for students who were not admitted into other institutions of higher education. Over the coming years it will be interesting to observe how teacher training programs change since this responsibility was delegated to the states in 1992.[61] The extent to which this training is incorporated into state universities will serve as a good indicator of its long-term viability. The states face serious economic difficulties because of the limited monies that the federal government allocates to each state for its educational expenses. The allotment for education represents an important part of the state budget, but for most states it only covers payroll and school operating expenses, leaving no funds to support teacher training programs and ongoing professional development.

It is essential to reformulate the vision and national imagery surrounding teaching and to increase its professionalization. Only in this way will a higher value be placed by society on the social function of the woman teacher. This reformulation also requires that there be the same number of women and men teachers within the schools, so that children do not grow up with their identities shaped by starkly unequal patterns of gender and power in society. Without these changes, Mexico will face a severe educational crisis in the immediate future brought on by the lack of highly qualified teachers. Placing a high value on the social function of the teacher is above all linked to the professionalization of the field, improved teachers' salaries, and better working conditions.

Teachers are clearly at the center of schooling. During the economic crisis of the 1980s, when teaching lost its ability to exercise its professional function because of the poverty-level wages teachers received in many areas of the country, teachers abandoned schools and looked for ways to supplement their income, which reduced their interest in professional development.

To value teachers properly, we must increase their professional autonomy and improve their working conditions. The lack of financing has had a pronounced negative effect on the school culture in Mexico. In particular, it is crucial to put an end to the perception that teaching is a "service vocation," because this stereotype helps to justify, among other things, low teaching salaries as well as the scant regard given to the woman teacher. It is assumed that her salary is a second income for the household, which does not take into account her responsibilities as a professional and her need for a salary that allows her to perform her professional duties.

In the case of Mexico, one cannot speak of a gradual feminization of the teaching profession, that is, a slow growth in the number of women teachers over the years. From the outset, a constant number of women—in relative numbers—have participated in education: in urban zones more than two-thirds of all teachers have been women since the beginning of the twentieth century. At the preschool level, the enrollment in training programs for preschool teachers at the *Escuela Nacional de Educadoras* (National School of Women Preschool Teachers) has always been restricted to women, and so this level of education has always been feminized.

As I indicated earlier, adjustments in economic policies have affected the conditions for rural and urban employment for women and men teachers. In the 1980s, during the economic crisis, when the wage of state employees dropped, there was a clear decline in the number of women teachers and an increase in men teachers, especially in rural zones where few other employment options exist.

An interesting phenomenon can be observed in Mexico starting in the 1990s. There has been a marked rise in the number of women in academic high schools and a corresponding increase in women at the university level. However, the increase in women's graduation rates from colleges and universities has not been accompanied by their greater presence in corresponding professional occupations. Women constitute a third of the labor force in Mexico but represent only a minority of professionals—only 3 percent. The educational pyramid displays a segmentation of functions differentiated by gender that proves how work performed by women is limited by their presumed maternal role and the differentiated socialization for boys and girls.

Within universities, women continue to be concentrated in majors that are traditionally considered female or have become increasingly feminized in recent years. Statistics from ANUIES[62] show that 50 percent of female university students are concentrated in five academic areas[63] (these statistics do not include teacher training institutions where women are a majority). In the 1970s, women were disproportionately represented in Normal Schools, and only made up a third of the students enrolled in academic high schools, which provide the most direct access to higher education.[64] Today, women are still concentrated in Normal Schools and in teacher education programs. Although women have obtained access to higher education, they are not participating in the job market. As a result, there is a high unemployment rate among educated women.

The relatively high percentages of women in traditionally female majors in higher education demonstrates that women's social and academic expectations have not changed, and their professional development continues to be limited to one option: the "feminine" careers. The State has promoted women's access to primary schools, secondary schools, and universities. But public policies have not intervened sufficiently to change the fact that the jobs women perform in the labor market are an extension of their domestic life: teaching, health, and taking care of others. The stable employment of women in the labor market and the social and economic devaluation of the teaching profession are reflected in the fact that young female university graduates with their diplomas are not willing to enter the teaching profession, which has become a badly paid occupation held by those of a lower socioeconomic status without a university education. The school system and school culture continue to educate children in traditional gender roles. Statistics have clearly shown that educational access does not change traditional gender patterns. Likewise, the academic expectations of women are determined by cultural norms and patterns of femininity that channel the professional expectations of women toward the so-called "feminine" professions.

By Way of Conclusion

At the beginning of the previous century, the earliest feminist teachers fought for coeducation and improvements in the labor conditions of women in teaching. Today, however, in the twenty-first century, we find that teachers are undermined by their economic and professional situation.

In Mexico, the professionalization of teachers is linked to a new conception of the teaching function and a change in the teaching culture. A new culture is necessary that reflects on and questions Gabriela Mistral's idea of teaching as a service vocation. As a "service vocation" there is no need for professionalization or ongoing professional development, nor for improved working conditions and salary. One objective of current educational policies must be to improve the teaching practice and its effectiveness in the classroom. To achieve these objectives, Mexico must reassess the initial training that most teachers receive. As long as teacher training programs continue to operate in a parallel but second-class track to university studies, it will be impossible to attract candidates with a high educational level to the profession.

In the United States and in various European countries, teaching is a highly feminized profession. This does not, however, preclude a middle-class salary and ample opportunities for an ongoing professional education for women teachers. In Mexico, unlike Europe and the United States, teachers have not been given the educational opportunities that permit them to teach with autonomy and professionalism. Another important difference is that teachers' unions in the United States and Europe have fought for the professionalization of teachers in order to secure better salaries and greater autonomy, while in Mexico the national union has sought to control teachers to benefit the group in power linked to the political system. Professionalization, in this case, would have to be linked to a redefinition of the role of the union.

Finally, women teachers in Mexico must take the initiative in order to increase their value as professionals. They must be the first to demand better educational opportunities, in order to achieve a higher level of efficiency and an increased quality of public education in Mexico. Otherwise, women teachers—and all teachers in general—run the risk that teaching will continue to be devalued, that schools will be staffed by teachers without credentials, and that new technologies will replace teachers instead of performing their appropriate role, which is to strengthen and increase the flexibility of teaching and learning.

Archival Sources

Documento del Archivo de la Secretaría de Educación Pública Escuela Industrial "Gabriela Mistral": Finalidades de la escuela, Plan de Estudios, Reglamento Interior de la Escuela" (México: Secretaría de Educación Pública, n.d.).

El Maestro Mexicano (SEP, 1944–1958).

Notes

An earlier version of this essay was presented at the First International Congress on the Processes of Feminization, El Colegio de San Luis, San Luis Potosí, Mexico, February 21, 22, and 23, 2001.

1. See chapter by Morgade and chapter by Molina in this volume.

2. Inter-American Development Bank, *Development Beyond Economics* (Washington, DC: IADB, 2000). See Dan C. Lortie, *Schoolteacher: A Sociological Study* (Chicago, IL: University of Chicago Press, 1975).

3. For example, in the case of Mexico City, 73 percent of teachers in the public education system are women. See Regina Cortina, "La educación de la mujer en Latinoamérica. La profesión de la enseñanza en México," in *Estudios de El Colegio de México* XIII, 39 (Sept–December, 1995): 601. Table 1 "La participación de la mujer en el sector educativo en la ciudad de México." In elementary schools in Mexico City, 79.5 percent of teachers are women. See María de Ibarrola, Gilberto Sillva Ruiz, and Adrián Castelán Cedillo (México: Fundación SNTE para la Cultura del Maestro Mexicano, AC, 1997), p. 33. According to statistics in the *IX Censo general de población y vivienda 1990*, prepared by the INEGI (Instituto Nacional de Estadística, Geografía e Informática), of those in the population who identified themselves as "educational workers," 60.3 percent were women. In Ibarrola, Sillva Ruiz and Castelán, Mexico: Fundacion SNTE para la Culture del Maestro Mexicano, p. 18.

4. Current statistics on the gender and age of teachers, the school calendar, and the hours of instruction required in each country by subject area can be found in the annual report *Education at a Glance: OECD Indicator* published by the OECD (Organization for Economic Co-operation and Development). This information is also available at OECD's website: www.oecd.org.

5. For a study on the entrance of rural women into teaching, see Mary Kay Vaughan, "Women School Teachers in the Mexican Revolution: The Story of Reyna's Braids," *Journal of Women's History* 2, 1 (1990): 143–167.

6. The first two feminist congresses in México were organized in the state of Yucatán in 1916. A wealthy state due to its henequen production, Yucatán opened its educational institutions to women and under two

socialist governors encouraged the participation of women in society. See chapter 2 and 3 in Shirlene Ann Soto, *The Mexican Woman: A Study of Her Participation in the Revolution* (Palo Alto, CA: R&E Research Associates, 1979).

7. See Anna Macías, *Against All Odds: The Feminist Movement in Mexico to 1940* (Westport, CT: Greenwood Press, 1982).

8. Ibid., p. xv.

9. See Alaide Foppa, "The First Feminist Congress in Mexico, 1916," *Archives*, 5, 1 (1979): 192–199.

10. See Regina Cortina, "Gender and Power in the Teachers' Union of Mexico," *Mexican Studies/Estudios Mexicanos* 6 (Summer 1990): 241–262; and Regina Cortina, "Women as Leaders in Mexican Education," *Comparative Education Review* 33 (August 1989): 357–376.

11. Mistral, born in Chile in 1889, was baptized Lucila Godoy Alcayaga.

12. She was awarded the Nobel Prize for Literature in 1945.

13. Document "Escuela Industrial 'Gabriela Mistral': Finalidades de la Escuela, Plan de Estudios, Reglamento Interior de la Escuela" (México: Secretaría de Educación Pública, n.d.).

14. Ibid., p. 3.

15. Gabriela Mistral, *Lecturas para Mujeres* (México: Editorial Porrúa, [1923]1980, 6th ed.).

16. For an account of the Educational Plan of 1867 under the Juárez administration, see chapter 8 in Ernesto Meneses Morales, *Tendencias Educativas Oficiales en México, 1821–1911* (México: Porrúa, 1983).

17. Josefina Zoraida Vázquez, "La educación de la mujer en México en los siglos XVIII y XIX," *Diálogos* (Abril 1981): 10–16.

18. See Andrés Molina Enríquez, *Los grandes problemas nacionales [1909]* (México: Ediciones Era, 1978).

19. See Jaime Torres Bodet, *Memorias* (México: Porrúa, 1981) in which he recounts the public's critical reaction to Vasconcelos's book project.

20. Roque Esteban Scarpa, *Gabriela Mistral, Magisterio y Niño* (Santiago de Chile: Editorial Andrés Bello, 1979), p. 172.

21. *El Maestro Mexicano* [The Mexican Teacher] (SEP, 1944–1958).

22. Cited in the prologue by Jaime Quezada in *Gabriela Mistral: Escritos Políticos* (México: FCE, 1994), p. 15.

23. A complete biography can be found in Palma Guillén de Nicolau, "Introducción," to Gabriela Mistral, *Desolación, Ternura, Tala y Lagar* (México Porrúa, 1992).

24. For a discussion of the influence of Froebel's ideas, particularly the description of teaching "as a conscious and articulated version of mothering," see, Carolyn Steedman, " 'The Mother Made Conscious': The Historical Development of a Primary School Pedagogy," *History Workshop: Journal of Socialist and Feminist Historians*, 20 (1985): 149–163.

25. Jaime Quezada, *Gabriela Mistral: Escritos políticos*. See essays entitled "Una nueva organización del trabajo," 1 and 2, pp. 253–260.

26. Document "Escuela Industrial 'Gabriela Mistral': Finalidades de la escuela, Plan de Estudios, Reglamento Interior de la Escuela" (México: Secretaría de Educación Pública, n.d.), see p. II.

27. Palma Guillén de Nicolau, *Gabriela Mistral (1922–1924)*, prologue to *Lecturas para Mujeres*, pp. ix–x.

28. *Lecturas para Mujeres*, p. 13.

29. Quezada, *Gabriela Mistral*, p. 13.

30. Roque Esteban Scarpa, *Gabriela Mistral, Magisterio y Niño* (Santiago de Chile: Editorial Andrés Bello, 1979), p. 198.

31. Alvaro Valenzuela Fuenzalida, *La vocación vertical: El pensamiento de Gabriela Mistral sobre su oficio pedagógico* (Valparaíso, Chile: Ediciones Universitarias de Valparaíso, 1992).

32. The model of maternal teachers was also used to bring in women to public education in Spain, see Sonsoles San Román, *Las primeras maestros: Los orígenes del proceso de feminización en España* (Madrid: Ariel, 1998).

33. Gabriela Mistral, *La vocación vertical*, p. 78.

34. Gabriela Mistral, *Magisterio y Niño*, pp. 220–221.

35. Gabriela Mistral, *Magisterio y Niño*, p. 249. This text is part of a speech given in Chile in 1916 or 1917 entitled "Conferencia para maestros; El cultivo del amor patrio." The similarities in the thinking of Mistral and the ideas of Vasconcelos regarding the function of a teacher can be seen in this document.

36. Secretaría de Educación Pública, Estadísticas Históricas del Sistema Educativo Nacional, August 2004.

37. Regina Cortina (ed.), *Líderes y Construcción de Poder: Las maestras y el SNTE* [Leaders and Construction of Power: Women Teachers and the SNTE] (México: Santillana, 2003).

38. Gordillo is also called by her followers "the moral leader of the teachers."

39. A position especially created for her in March of 2004.

40. Sabina Berman and Denise Maerker, *Mujeres y Poder* (México: Raya en el Agua, 2000), p. 75.

41. A publication of the executive committee of the SNTE.

42. The ISSSTE is the social security system for state employees.

43. *The Beijing Declaration and Platform for Action* (New York: United Nations, 1996).

44. CEDAW, Convention on the Elimination of All Forms of Discrimination against Women, adopted in 1979 by the UN General Assembly.

45. Regina Cortina and Nelly Stromquist (eds.), *Distant Alliances: Promoting Education for Girls and Women in Latin America* (New York: Routledge, 2000).

46. CERFAMI, "Reconocer y superar el sexismo en el proceso educativo" (Medellín: Colombia, 1999) p. 54.

47. Idid., p. 76.

48. Carmen Colazo, "Public Policies on Gender and Education in Paraguay: The Project for Equal Opportunities," in Cortina and Stromquist, *Distant Alliances*, pp. 13–27.

49. Rosa María González, *Construyendo la diversidad: Nuevas orientaciones en género y educación* (México: Porrúa, 2000).

50. For an analysis of this experience, see Malú Valenzuela and Gómez Gallardo, "Other Ways to be Teachers, Mothers and Fathers: An Alternative Education for Gender Equity for Girls and Boys in Preschools," in Cortina and Stromquist, *Distant Alliances*, pp. 103–118.

51. Sylvia Schmelkes, et al., *The Quality of Primary Education: A Case Study of Puebla, México* (París: UNESCO, IIPE, 1996).

52. Ibid., pp. 80.

53. Ibid., pp. 68.

54. María de Ibarrola, "Education and Economic Growth: Creating a Culture of Education." Paper presented at the conference "Reforming Education in Latin America: The Second Wave of Reform" (Council of Foreign Relations, New York, March 20, 1996) Table 3: Mexican Teachers' Salaries, 1978–1990.

55. International Labor Organization, *The Impact of Structural Adjustment Policies on the Employment and Training of Teachers* (Geneva: ILO, 1995), p. 51.

56. Ibid.

57. María de Ibarrola et al., *¿Quiénes son nuestros profesores? Análisis del magisterio de educación primaria en la Ciudad de México 1995* (México: Fundación SNTE para la Cultura del Maestro Mexicano, AC, 1997), p. 34.

58. Ibid., p. 79.

59. Ibid., p. 131.

60. Juan Carlos Tedesco, "Los docentes y la reforma educativa," in: Hernando Gómez Buendía (ed.), *Educación la agenda del siglo XXI: Hacia a un desarrollo humano* (PNUD, Colombia: TM Editores, 1998), pp. 254–256.

61. Through the National Agreement for the Modernization of Basic and Normal Education.

62. Asociación Nacional de Universidades e Institutos de Educación Superior.

63. ANUIES Statistical Report, 2003. Data from 2002 is reported.

64. See Cortina, *Líderes y Construcción del Poder*.

CHAPTER 5

Educational Policies and Gender: An Assessment of the 1990s in Brazil

Fúlvia Rosemberg

INTRODUCTION

In the decade of the 1990s important changes to Brazilian educational policies were put into effect in the aftermath of the Constitution of 1988 and new international configurations. These changes translated into reforms that increased educational access, adjusted the students' ages by grade level,[1] and sought to improve the quality of education at the same time that the use of public funds was restricted. These reforms were not exclusive to Brazil and Latin America; rather, they represent part of an international movement that conceives of education as a fundamental strategy for the reduction of socioeconomic inequalities at the national and international level.

The most noteworthy reforms in the 1990s were the promulgation of the new National Education Law[2] (law no. 9394, passed in December, 1996); the formulation of National Curriculum Parameters for different levels of education; the introduction of the National Systems of Evaluation for elementary and secondary education (SAEB) and higher education (ENC); implementation of policies concerning the evaluation and distribution of textbooks; and rules governing the financing of education, especially through the law that created the Fund for the Development and Support of Basic Education and Valorization of Teaching (FUNDEF).[3] However, if an evaluation were done of the impact these reforms have had on the democratization of education and on the outcomes that could be attributed to schooling, no consensus would be reached. Progressive researchers and educators have called attention to the fact that the survival of the educational sector is tied to the laws of the market, and they emphasize the strong economic basis of the reforms. The World Forum on Education, held in Porto Alegre, Brazil in October 2001, gathered together a significant number of opponents of these hegemonic tendencies, which grew out of the "Washington Consensus."[4]

In this context, education has a prominent place on the world agenda for development. Numerous world conferences on education were held in the 1990s, under the auspices of the United Nations and its affiliated organizations: UNESCO, UNICEF, and the World Bank. Brazil was one of the nations that signed the agreement for the Education for All Conference /EFA (held in Jomtien and Dakar) and was an active participant in the EFA-9 group, which gathers together the nine most populated countries in the developing world. Brazil also signed international agreements relating to the equality of educational opportunities for women and men which derived from multilateral agreements signed at various international conferences during the 1990s, also sponsored by the United Nations. These agreements emphasize the importance of education for women and justify it as one of the most important strategies to combat social inequalities and underdevelopment.

Various official documents in Brazil have reiterated that the Brazilian educational system is free of gender discrimination and bias against women, as can be seen in the following excerpt from the national report that was prepared for the Dakar Conference:

> The guidelines established in Amman and Islamabad had already been incorporated in the educational policy of the Brazilian government, especially those referring to the importance given to the training, status, remuneration and motivation of teachers. Additionally, the recommendations of the Fifth International Conference on Adult Education (Hamburg, 1997) constitute a central concern of the National Plan of Education. Of the objectives and goals established at the EFA-9 conferences, only those relating to the prioritizing of education for women and girls were not incorporated in Brazil, where this problem is not found. Gross enrollment rates, school completion rates, and average years of schooling are higher in the female population than in the male population. As long as this trend continues, the concern over gender issues in Brazil will have to be stated in opposite terms.[5]

That is, the Brazilian government, along with multilateral organizations and hegemonic groups from the women's movement, claim that gender opportunities in the Brazilian educational system are equal, having reduced their argument to indicators of access and how many years children stay in school. They forget other aspects of education that reflect, sustain, and create gender inequalities: the educational system that participates in the formation of a citizen; the educational system as a job and consumer market; the educational system that places obstacles in the path of women and men from lower social classes who want to continue their schooling, such as, non-whites, those from rural zones, from families with few resources, or from the North and Northeast regions.

The purpose of this chapter is to describe the trends and indicate the advances and gaps that exist in educational policies from the perspective of gender equity. I will attempt, in this way, to make a contribution to the broad discussion about the hegemonic agenda shared by the Cardoso administration (1995–2002), multilateral organizations, and important segments of the diverse women's and feminist movements,[6] both national and international.

In studies that I have carried out in recent years, I have shown that these social actors, backed by an often incomplete situational analysis and by a theoretical model that is far from adequate for the educational system, have poorly served an agenda that calls for equality of gender opportunities in education. It is even possible to say (guardedly, however) that the present agenda shared by multilateral organizations, governments (including Brazil), the feminist and women's movements, and gender/women's studies, simplifies or impoverishes our understanding of the process of gender domination in education. I will offer a profile, then, of the education of men and women in Brazil, showing where I find its weak points. In this way, I will question the hegemonic agenda.

This chapter will use the following terms: sex, to refer to men and women on the basis of their civil identity, and constituting, therefore, a disjunctive variable; gender, to refer to symbolic constructions with respect to masculine and feminine in our society, which protect and sustain the ideology that the masculine is superior to the feminine; color, to refer to the self-classification of an individual, which corresponds to one of five alternatives of denomination proposed by the *Instituto Brasileiro de Geografia e Estatística* (IBGE)—white, black, brown, indigenous, and yellow—taking into account that the denomination "black race" or "blacks" is reserved for the subgroup composed of individuals who declare themselves to be black or brown.[7]

Part of this chapter is based on the macro-analysis of data, and privileges the information collected and published by the

IBGE in its National Household Surveys, henceforth referred to as PNADs.[8]

EDUCATIONAL STATISTICS

In Brazil, three institutions produce statistics on education: (1) the Ministry of Education and Sports (MEC),[9] or through the National Institute of Statistics and Educational Research (INEP)[10] and the Department of Education for each state, whose data is mainly collected at educational facilities; (2) the IBGE, which uses the home as its principal source of data; and (3) the Ministry of Labor, through its Annual Reports of Social Statistics (RAIS),[11] which also provides information on the teachers who work in the formal labor market, and which collects data from the schools.

Each of these agencies has its own specific data-collecting methods and has defined its own population; thus their findings do not necessarily coincide. For example, while IBGE statistics refer to students, those from MEC refer to number of students enrolled; and the number of students enrolled at any given time throughout the school year is not always the same. Moreover, the variables selected to characterize one group and another may also vary.

Brazilian educational statistics have improved, especially during the last federal administration. In recent years, international organizations (OECD, UNESCO, UNICEF) and the feminist and women's movements have stressed the importance of providing a statistical breakdown by sex. This practice, which is very common inside the IBGE and MEC, has been expanded: for example, statistics on sex and color/race were also included in recent surveys measuring students' performance, such as in the System of Evaluation of Basic Education (SAEB) and the National Examination of Courses (ENC).

For this reason, the country has at its disposal a rich and complex store of statistical information on literacy programs, instruction, school enrollments, matriculations, percentage of students who finish each academic year, types and quality of schools, pass/failure rates, results on national tests, professional teacher training, and the number of teachers as a percentage of the economically active population. However, the dissemination of statistics broken down by sex, specifically those gathered through the Educational Census performed by the MEC/INEP, are somewhat more precarious, and reflect only a small part of the wealth of statistical data that has been collected.

This information, I fear, is processed and released only under special circumstances (e.g., on International Women's Day). One must keep in mind MEC's position that Brazil has overcome gender inequalities in the educational sphere.[12] The lack of systematic analysis and poor diffusion of statistics that consider the sex of the respondent is particularly notorious in childhood education, in higher education, in master's and doctoral programs, and in innovative programs that are intended to improve the students' advancement.[13] Therefore, statistics on the different levels[14] of the formal school system, which might reveal new processes of discrimination, are scarcely disseminated, and if available, do not consider sex and color/race. It should also be noted that there is little information on education for men and women who belong to indigenous or gypsy communities.

However, when considered from a gender perspective, the most important trends in the systematic collection, diffusion, and analysis of educational data are centered mainly on the school, or more specifically, the teaching staff and students. Bear in mind that even formal education goes beyond school walls, especially at present with the spectacular spread of new educational technologies utilizing computers and television, along with the wide availability of traditional didactic materials, such as books and toys. The educational system is not an island: civil construction, transportation, equipment, clothing, and didactic materials of all types, are some of the economic sectors that affect and are affected by education, especially after its expansion. Also note that statistics about the educational market on the national and international level keep the dynamic of gender relations shrouded in mystery. Questions related to the central role of education in the world agenda, the processes through which formal and informal educational systems are expanded, the globalization of the education market, the diffusion of educational technologies and new gender dynamics (and, one might add, those of class and race) in the job market and in the domestic space, have not been formulated—I will argue—because statistics are not collected using these lenses. The reverse is also true. Without new questions, it is difficult to collect new kinds of data, and what we find is a chaotic research agenda and policymaking that was put in place prior to the expansion of educational systems, the changes in the world economy, and the spread of new educational technologies. In my opinion, these constitute crucial points to be incorporated into a new program of research studies on education and gender.

In their absence, I will attempt to analyze existing educational indicators about the men and women in the Brazilian educational system based on data collected in the schools.

ILLITERACY AND LITERACY

Although the rate of Brazilian illiteracy has decreased, it remains high: 26.6 percent in 1985 and 15.7 percent in 1999 among the population age five and older; 21.5 percent in 1985 and 13.0 percent in 1999 in the population age seven and older.[15]

If we compare the statistics gathered from the 1872 through the 1999 Censuses, rates of male and female illiteracy follow a parallel course that can be observed up to the 1940s. Until that decade, the rate of difference between male and female illiteracy rates was relatively high,[16] with greater female illiteracy. After 1940, both the illiteracy rate and the gap between male and female illiteracy decreases. Women's access to formal education, which intensified around 1940, contributed to a significant reduction in the percentage of illiterates in the country during the 1940s. This has been confirmed by other Brazilian researchers who, however, attribute it exclusively to the expansion of the formal educational system.[17]

The profile of the illiterate female is almost identical to the male. They are women and men from the lowest income strata: blacks living in rural zones and in the Northeast, who face the greatest barriers to achieving literacy.[18] The greater access of women to education and their higher utilization of schools are reflected in the literacy rate. At present, the percentage of literate women exceeds that of men: 84.7 percent among females and 83.9 percent among males in the population age five and older.[19]

The difference in literacy rates for males and females, broken down by age, is hardly perceptible: among the age group 15 to 19 (the group with the highest literacy in the country), women have greater literacy than men (97.3 percent and 94.7 percent, respectively); in ages 50 and over (the group with the lowest literacy overall)[20] the literacy rate for men is higher than that for women: 73.1 percent and 68.0 percent, respectively.[21]

The higher rates of literacy in the female population are a result of girls' greater access to schools and higher school utilization rates. The higher rates of literacy among the oldest men are a legacy of the past, when women's access to school was more restricted, and there was a lack of literacy programs for older women in the population.

Contrary to common belief, the profile of Brazilian illiteracy by color/race is significantly different from the profile by sex: the largest group of illiterates is historically constituted by blacks (blacks/browns), in relative as well as absolute numbers. It is time for researchers, administrators, and feminist activists to realize that the processes of race domination do not follow the same logic as gender domination, nor are they simultaneous, whether viewed from the lens of social history or from an individual's life trajectory.

EDUCATIONAL INDICATORS FOR MEN AND WOMEN

The difference between men and women in the formal Brazilian educational system does not show striking variation depending on age and grade level in school. They are more apparent over the course of an individual's school trajectory rather than in any specific barriers to school access.

Women represent 51.3 percent of the overall population age five and over and 50.5 percent of all students from that age group.[22] The rate of school enrollment for men is therefore slightly higher than for women (35.6 percent and 32.5 percent, respectively). This means that the drop-out rate for the Brazilian population age five and older is, in percentages, not as high among men. Starting in 1996 women exceeded men in average years of schooling and they continued to show more visible gains in the last decade. However, the average years of schooling for both sexes continues to be low (see table 5.1).

The apparent contradiction between these two indicators—the rate of gross enrollment and average years of schooling—and their changes

Table 5.1 Average years of study in the population age five years and over, by sex—Brazil

Sex	Years				
	1960	1970	1980	1990	1996
Men	2.4	2.5	3.3	5.1	5.7
Women	1.9	2.4	3.2	4.9	6.0
Total		2.4	3.3	5.0	5.9

Note: The rural population from the North is not included in 1990 and 1996.

Source: 1960, 1970, and 1980 Census Demographics; 1990 and 1996 PNADs (cited in IPEA/PNUD).

in the 1990s is caused by the different school trajectories of men and women. In fact, women's advancement in school is more regular than men's, creating an educational pyramid whose apex is slightly flatter, and is therefore somewhat less selective, a tendency that was accentuated during the 1990s.

For both sexes, school advancement encounters a similar bottleneck because students fail or drop out. However, advancement for men is more problematic. On the average, a Brazilian man or woman needs 10.4 years to complete the eighth grade of elementary school, representing an efficiency rate of 0.78, which is reflected in the age–grade gap (the difference between the ideal age and the actual age of the student at a specified grade level) see figure 5.1. This can be explained not only by a delayed entry into the educational system, but also by failures, or departures and returns.

The age–grade gap is less pronounced for women than men in the two racial groups. The interracial comparison shows that blacks and browns display a larger gap than whites, and that the age–grade gap for black men is greater than that for black women, just as the

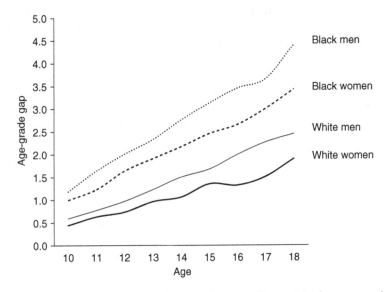

Figure 5.1 Average age–grade gap for ages 10 to 18, by race and sex—Brazil, 1999

* Average age–grade gap is defined as: Age minus grade minus 7 (years).

Source: 1999 PNAD. The rural population from the North region is not included. Statistics processed by Sergei Soares (IPEA).

age–grade gap for white men is greater than that for white women. Despite this pattern (which researchers are aware of), an "associativist" model of Brazilian educational inequalities has circulated within the country and outside, which assumes that color and sex are linearly connected, and believes, erroneously, that black girls and women are the ones who will exhibit the worst educational indicators. The empirical data does not support this model. Several research investigations have shown that black women, across nearly all age groups, display better educational indicators than black men, in the same way that white women display better educational indicators than white men as is shown in table 5.2.

The differences in school advancement between men and women, white and black, create distinct educational pyramids: the pyramid for white men and women, is slightly flatter (and therefore less exclusive) than the one for black men and women; moreover, the pyramid for women, both white and black, is not as flat as the pyramid for men in their corresponding racial group.

Consequently, in any given school year, more women than men finish elementary, secondary school, and undergraduate levels. As reported by the then-president of the INEP in 2000,

> Among those students in the final year of elementary education, 53.6 percent are female and 46.4 percent are male. In secondary education the same phenomenon is observed: 58.3 percent of those who finish are female and 41.5 percent are male. The female hegemony is even more accentuated in universities, with women comprising 61.4% of those who finish.[23]

Table 5.2 Sex and race distribution of students age five and over, by educational level—Brazil, 1999

Education Level	Men		Women	
	White	Black	White	Black
Preschool	9.4	10.3	8.9	9.3
Elementary	66.3	76.6	61.7	73.9
Secondary school	16.1	11.2	19.2	14.1
Higher education	8.2	1.9	10.2	2.7

Note: The rural population from the North is not included.

Source: 1999 PNAD cited in 2001 CNDM (Conselho Nacional dos Direitos da Mulher).

In summary, the slightly higher proportion of men in the student population, indicated at the beginning of this section, can be explained by the fact that they remain longer in the school system to cover the same trajectory as women. The patterns in the trajectories of men and women, whites and blacks, are different for the pre-school, elementary, and secondary school levels, and are reflected in the age of students at the beginning and end of undergraduate studies. However, while we find a higher proportion of younger women who begin and complete their education, we also find a greater proportion of older women (age 25 and over) who begin programs of higher education. This seems to indicate a greater return to school on the part of women who, due to the contingencies of life, had earlier abandoned their studies. This trend is also evident when we analyze students in *supletivo* programs,[24] in which there is a slightly higher percentage of women: women comprise 50.4 percent of students in *supletivo* programs for elementary education and 51.4 percent for secondary education.[25] This might be a result of the overrepresentation of females in the indicated age range, or to an active search for education on the part of women.

Researchers, feminist activists and Brazilian administrators seldom ask themselves what impact educational expansion and contemporary educational reforms have had on the schooling of men and women. I have been able to answer this question in part, yet admit that this chapter is a preliminary exercise that seeks to add complexity to the optimist or pessimistic simplicity of many who have written on women's education from a gender perspective. For this analysis I have chosen six themes: the expansion of the school system; the specialization within fields of study; the teaching profession; curricular reforms; the politics of school textbooks; and early childhood care and education.

The Expansion of the School System

One of the fundamental goals of the educational sector during the Cardoso administration (1994–2003) was the expansion of the school system and the broadening of its offerings. To evaluate the impact of school expansion, three indicators were selected: literacy rates, gross enrollment rates, and the percentage of individuals at different educational levels. I then calculated their rate of growth and compared it to the rate of increase in the overall population within the same age group. The need to control the rate of population growth is due to the unequal number of male/female births as well as different life expectancies.

In Table 5.3, which reports the rate of enrollment growth (and the relative difference between 1985 and 1999), an interesting trend rarely mentioned in the literature can be observed: an increased rate of access to the educational system that is slightly more accentuated for men (51.4 percent for men and 50.2 percent for women between 1985 and 1999). Women, however, have increased their academic advancement within the educational system, especially at the under-graduate level (see rates of increase for undergraduate students and among individuals with at least 12 years of schooling).

To determine whether the Brazilian educational system has privileged one sex over the other, it is necessary to control the sex ratio (the percentage of women in the population that is represented by the gross enrollment rate). Between 1985 and 1999, the growth in the

Table 5.3 Rate of growth from 1985 to 1999 by selected educational indicators and by sex—Brazil

Indicators	Rate of growth*	
	Male	Female
Population (age 5 and over)		
Total	25.4	28.5
Urban	37.2	40.0
Rural	−5.1	−5.5
Literacy (age 5 and over)		
Total	43.3	48.2
Urban	49.4	54.1
Rural	19.8	23.0
Students (age 5 and over)		
Total	51.4	50.2
Preschool	84.1	78.4
Elementary School	34.9	28.9
Secondary School	167.2	159.9
Higher Education	69.9	123.4
Level of Education (10 years or more)		
Total	30.8	33.9
Population without education	−15.9	−18.8
and up to 1 year		
4 years	8.2	9.9
8 years	76.9	76.8
9 to 11 years	102.2	133.3
12 years and over	73.9	125.8

Note: The rural population from the North is not included.
*Percentage of growth: Total for 1999 − Total for 1985/Total number of minors × 100.

Source: 1985 and 1999 PNADs.

Brazilian female population age five and over exceeded male growth by 3 percentage points. This is a result of women's greater longevity, along with the premature death of teenage and young adult men, especially from violent causes. It is possible to construct hypotheses or explanatory models that link the gender ideology of our society, the educational system, and sexual differences with life expectancy. Although this would be desirable, it goes beyond the scope of this work.

This data suggests that the Brazilian educational system favors relatively greater access by men and relatively higher advancement by women. This trend, if it is confirmed in more detailed studies (those based on census data, for example), can be attributed to educational reforms, such as automatic promotions, accelerated programs and classes, and is associated with the observed overrepresentation of women in private universities (a phenomenon that deserves a more thorough investigation). That is, there are signs (which need to be studied further) that in contemporary Brazil, public educational policies are not egalitarian, nor do they privilege women, as had been claimed in the national report prepared for the Dakar conference that was cited earlier.

This analysis underscores the urgent need for data on matriculations in the different school networks and at the different levels (including high school) by sex. Although this information is collected by MEC/INEP it is not always processed and released. Likewise, data is needed that examines by their sex and color/race those students enrolled in "innovative" programs (like the advanced classes) or in other recent initiatives such as evening programs, which were intended to increase students' access to education and meet international goals. There are strong indicators that the "women" privileged by the accelerated classes and evening programs are adolescent black males. The questions raised are: What type of education is being offered so that students enter or remain in the system? How many advance? How do they advance? What do they learn? What effects can be expected in the labor market and in their private life? How will their exercise of citizenship be affected? What specializations will they choose?

Fields of Study for Women and Men

While women may encounter fewer barriers than men and have more energy to complete higher levels of education, there is still, however, a marked tendency within the Brazilian educational system toward sexual differentiation within fields of study. This means that once in school, women tend to take more preparatory courses and men tend to take more professional courses. At the university level there is still a certain polarization between literature and human and social sciences (more feminine) and exact and technological sciences (more masculine).

Three tendencies characterize the distribution of men and women in the different branches of education. First, differentiation between the sexes tends to occur as early as the school system allows it; second, it remains relatively constant throughout subsequent grades; and third, there are no signs that specialization by sex in academic fields will disappear, although it is decreasing. According to the Census of Professional Education (2000), women represent only 39.3 percent of students matriculated in professional education, that is, in courses oriented toward the job market. A segregation by sex within specific areas of professional education can also be seen.

For higher education, we have statistics from 1980 and 1999 (see table 5.4). The data, however, is not entirely comparable, insofar as the names and groupings of the specializations have changed over the course of the years, as well as the unit of analysis: students or graduates.

The data shows a marked change in the levels of segregation by sex across different subject areas. The expansion of higher education not only caused an increase in both women and men students, but it also increased their options among a greater selection of majors.

Table 5.4 Percentage of women among university students or graduates by subject area

	Third grade	
Subject area	1980 Students	1999 Graduates (diplomados)
Biological and health sciences	40.6	68.9
Exact and earth sciences, engineering, and technology	18.2	40.1
Agricultural sciences	9.7	37.9
Human and social sciences	53.8	63.3
Linguistics, literature and art	86.2	83.3

Note: The original data was grouped together and should be analyzed with care.

Source: 1980 Census, cited in Rosemberg, F; Pinto, R. P. Negrão, Esmeralda, V. *A educação da mulher*, São Paulo: Global, 1982; MEC/INEP (1998) cited in CNDM/Conselho Nacional dos Direitos da Mulher. Educational Indicators: educação. 2001, www.mj.gov.br/sedh/cndm.

In a recent study, the professional choices of 156 boys and 151 girls in the fourth grade of elementary school in the city of Belo Horizonte were compared to responses given in similar studies from previous years.[26] The researchers[27] found significant changes as well as important constants: a trend that continues unchanged among girls is the choice of a liberal and "caregiver" profession, but there is a decline in the girls who chose "teacher" and "entertainer." Among the boys, the current choices are, in first place, "athlete" and "entertainer," and far behind, in second place, the liberal professions (engineer and lawyer), which inverts a tendency that was observed many years ago when this issue was first investigated. In addition to the premature polarization in career choices, these results also point to academic trajectories of different lengths for boys and girls.

In Brazil, contemporary educational reforms introduced systematic and broad evaluations of school competency. The results of these evaluations are broken down by sex, as in developed countries such as the United States, Canada, and France. Women tend to achieve better results on language tests, and men outscore women on mathematics and science tests (table 5.5). These results, however, are not consistent over the course of an individual's passage through school and must be interpreted with care, for as we have already seen, in any given school year, the socioeconomic status, racial composition, and the age group of men and women is not the same.

To summarize: the Brazilian educational system exhibits a pattern that is similar to, but not the same as, the pattern in developed countries. On the one hand, we observe better school performance among

Table 5.5 Performance on the SAEB* by grade level, subject area, and sex—Brazil, 1999

Subject area	Grade level	Sex		
		Men	Women	Difference (W−M)
Portuguese Language	4th Elementary	167.26	174.74	+7.48
	8th Elementary	227.16	238.07	+10.91
	3rd Secondary school	260.36	271.06	+10.70
Mathematics	4th Elementary	181.26	181.12	−0.14
	8th Elementary	252.88	240.82	−12.6
	3rd Secondary school	289.37	274.42	−14.95

* *Sistemas Nacionais de Avaliação da educação básica.*

Source: SAEB Report 1999 (www.inep.gov.br, accessed 01/15/01, at 18:20).

women that can be attributed to the continuing male–female segregation in chosen fields of study. On the other hand, we have seen an interrupted and uneven school advancement among women and men from subordinated social and racial sectors.

By charting sexual differentiation by specific areas of study, it is possible to extrapolate educational policies, because the current configuration seems to result from intervening patterns of gender, family, religious and peer socialization, in addition to a strong segregation by sex in the job market. As a result, recommendations aimed at promoting heterodox professional education in school for men and women in a highly segregated labor context are of little use without altering other socializing institutions, as well as the labor market. In addition, the democratization of access to a quality education presents a challenge to contemporary educational policy-making. It will not be achieved without the professionalization of teaching, a female-gendered activity that is mainly performed by women.

The Teaching Profession

The educational system overall continues to be a female occupation: whether working as teachers, or in other jobs (both white and blue collar), women constitute more than 80 percent of the workforce in the educational sector. Teaching continues to be one of the principal niches for women's insertion into the labor market: in 1980 teaching represented 8 percent of the Economically Active Population that was female; in 1991 it was 12 percent.

A comparison between the 1980 and 1991 censuses shows a slight decrease (−1.2 percent) in the percentage of women in teaching: In 1980, 86.5 percent of teachers were women; in 1991, 85.4 percent, We observe also a slight redistribution in educational levels: there was a small drop in the percentage of women teaching at pre-school and elementary levels and an increase at the secondary and university levels.

However, the slight changes between 1980 and 1991 did not alter the pyramid: men are under-represented in the grades for children and adolescents, and are over-represented in higher education, indicating strong gender discrimination. Elementary school teachers for Grades 1–4 are only required to have a secondary school education, and their salaries are markedly lower than salaries for teachers at the university level.

According to the 1998 RAIS report,[28] preschool education in Brazil continues to be the occupation that has the greatest percentage of women (94.8 percent) (see table 5.6). However, among elementary

Table 5.6 Percentage of women in teaching (fields of study) by year—Brazil, 1988 and 1998

	1988		1998	
Specialization	No.	%	No.	%
Preschool Education	69941	93.6	121355	94.8
Special Education	5746	81.6	19169	88.0
Elementary Education	133782	81.1	201088	80.2
Secondary School Education	369602	72.4	449447	72.2
Pedagogical Disciplines	20674	44.8	17328	58.8
Professional Education	35758	46.0	59132	56.0
General Secondary School Education	40339	37.3	49671	43.8
Biology and Medicine	18510	37.4	13652	43.6
Human Sciences	16990	43.1	17846	41.5
Physics and Chemistry	3994	31.2	2791	38.3
Mathematics, Statistics	4978	30.7	5229	36.1
Engineering and Architecture	10016	15.3	6596	23.5
Administration, Economics, and Accounting	11396	19.5	8745	23.2

Source: RAIS (1988 and 1998) cited in Bruschini and Lombardi (www.fcc.org.br, accessed 01/15/01 at 15:00).

school teachers there was a small decrease in female participation: 94 percent in 1978 to 91 percent in 1999. Codo[29] points out that the tendency toward a slightly higher presence of men among workers in the educational sector has intensified in the last few years as a result of the progressive reduction of teaching positions in the elementary school sector and of increases in the secondary school sector.

For the teachers working in the formal labor market, the difference between men's and women's education background and salary suggests that, in this broad professional field, teachers with very distinct profiles coexist.

The majority of women teachers finished secondary school (53.9 percent) and the majority of men teachers continued studying until they completed university (62.9 percent). This could justify, in part, the difference in the mean salaries (7.41 and 5.08 for men and women, respectively). However, at every level of instruction, the average salary for men is greater, underscoring the fact that there are different destinies for individuals who do not follow the typical careers for their sex: an invisible ceiling for women in male professions and an invisible escalator for men in female professions. However, what also defines the trajectory of male and female teachers is the age of the students. Women teach young children; men teach adults. Contemporary educational reforms have not changed the essence of these structural inequalities.

Several studies indicate that women, not men, face more unfavorable conditions in their work as teachers: Codo refers to the large number of women and men teachers affected by work-related illnesses; more women teachers than men work in rural school zones; and the number of teachers who are *leigos* (without the required professional qualifications) is higher for women than men.

Salaries for teachers, especially preschool and elementary teachers, continue to be very low, despite increased subsidies by FUNDEF.[30] According to the Brazilian report at the EFA Conference in Dakar, between December 1997 and August 1998, there was an increase of 12.9 percent in salaries for elementary teachers, reaching 49.6 percent in the municipal school system in the Northeast. This increase is relative, as it is necessary to consider the hours worked and the student/teacher ratio. That is, it is necessary to verify whether the salary increase corresponded to a greater valorization of the profession, or to a greater amount of work. There is evidence, for example, that the number of students per teacher increased at the elementary level.

The *Censo do Professor 1997*[31] revealed, in turn, discriminatory practices against women in teaching, although little data has been published that includes a breakdown by sex. In 1997 a census recount was done of 1,617,611 elementary school teachers, of whom 85.7 percent were women. Their median salary was $378.67 reales (Brazilian currency) per month, an amount that varied tremendously depending on the school network in which they worked and the region as is shown in table 5.7. Significantly, in elementary school education, the higher the participation of women on the teaching

Table 5.7 Average teacher's salary in reales (Brazilian currency) by geographic region and grade level—Brazil, 1997

		Grade level		
Region	Preschool or literacy	Elementary 1st to 4th	Elementary 5th to 8th	Secondary school
Brazil	419.48	425.60	605.41	700.19
North	322.01	360.77	586.37	735.46
Northeast	195.00	231.17	372.41	507.82
Southeast	587.00	613.97	738.57	772.09
South	464.96	460.12	594.44	683.03
Centralwest	573.64	447.55	584.20	701.79

Source: MEC/INEP (1997).

staff, the lower the average salary. Also, there is a significant variation in thé average salary depending on the grade of instruction (the lower the age of the students, the lower the salary of the teacher).

One may conclude, then, that the persistence of such low salaries for early childhood and elementary school teachers (who occupy 3.6 percent of the female workforce in the formal economy) contributes to the great salary differentials for Brazilian women workers. Without a doubt, this is a pattern of gender discrimination that still exists in the Brazilian educational system, attracting little or no attention within the international agenda for education in the 1990s. An important niche in the female labor market, the educational system, both public and private, produces gender discrimination by discrediting a profession that is considered feminine and chiefly exercised by women. In addition, it accepts, sustains and creates a deep inequality by age: small children are penalized because they are educated by less qualified teachers who receive the worst salaries.

Teaching is an exemplary case, although not an isolated one, of the gender discrimination that still persists in the education labor market. Even with the significant increase in the educational levels of women, which improved the educational level of the Economically Active Population overall, salaries for women continue to be lower than those for men with the same educational level. In addition, as I stated earlier, it is necessary to examine the full scope of the educational market and analyze the salary differentials between men and women. Based on this data, we could widen our reflections about those who have borne the costs of educational expansion, given the means with which it has been carried out: teachers, and poor families and their children.

EDUCATIONAL REFORMS AND THE AGENDA OF THE FEMINIST AND WOMEN'S MOVEMENTS

The National Program of Human Rights and the educational reforms of the 1990s once again took up three former topics on the agenda of the feminist and women's movements concerning education that elicited different reactions: the inclusion of sex education in the school curriculum; the battle against sexism in the school curriculum, especially in textbooks; and the expansion of preschool education as a means to provide childcare and education to the children of working mothers.

Changes to the Curriculum

Shortly before the National Program of Human Rights was announced on March 8, 1996, MEC and the Ministry of Justice, through the National Advisory Board on the Rights of Women[32] signed a protocol of equality between men and women. The protocol states that:

> The Ministries of Justice and Education agree to collaborate in order to guarantee that the educative process is constituted as an efficient instrument to combat all forms of discrimination against women, and to promote the recognition of their dignity, equality and their full citizenship. The agreement will be implemented through two courses of action to be developed by MEC: I) to incorporate themes promoting the recognition of equal rights between men and women into the curriculum of the Televised School, II) and in the choice of school textbooks to be adopted for the first and second grades, to use as one of the criteria non-discriminatory content with respect to women.[33]

The first initiative has yet to be achieved. However, MEC, in a certain sense, went beyond the limits of this agreement by including the topic Sex Education among the subjects in the National Curricular Parameters (PCNs)[34] for elementary school education. In the chapter on Sex Education, three pages are devoted to Gender Relations, and this issue is also addressed in passing in two other topics (for example, in Work and Consumption).

An analysis of the sub-topics within the chapter on Sex Education remains to be done. The declared commitment to educate the citizenry and to respect diversity can be seen in the Introduction to the National Curricular Parameters, which affirms that elementary school education should prepare students to confront "any discrimination based on differences in culture, social class, beliefs, sex, ethnic background or other individual and social characteristics."[35]

Despite criticisms—for example, its tendency to emphasize the heterosexual model of the family and sexuality—the PCNs present, for the first time at a national level, a secular vision of sexual education for elementary school students. Regrettably, it has not been adopted for the other educational levels.

The politics of textbooks was another topic under discussion. For Beisiegel, the Department of Elementary School Education of the Ministry of Education found that the National Textbook Program (PNLD)[36] was one of the principal means to implement the guidelines on democratic education that had already been established in the PCNs.[37]

The fight against sexism in textbooks was one of the issues most frequently raised by the Brazilian feminist and women's movements in the 1970s and 1980s, and it generated a number of research studies. Esmeralda Negrão and Tina Amado[38] have found forty-four investigative studies that directly or indirectly examine this theme. These studies demonstrate that in the period up to 1989, school textbooks tended to recreate, on a consistent basis, stereotyped images of men and women according to the traditional view of gender: the public sphere is reserved for men (work, leisure time activities, external space, activity, wealth, and power) and the private sphere for women (reproduction, internal space, passivity). Some of the studies examined in Negrão and Amado's report found that these stereotypes tended to be more pronounced when they dealt with individuals from other racial groups.

Research studies on stereotypes in textbooks exposed and denounced gender discrimination; however, they tended only to explain traditional and passive behavior among women, rarely contributing to an increased knowledge about gender discrimination in schools. Many of the studies were used to help sensitize teachers to what has lately been denounced in Brazil as differentiated education.

The widespread use of these research findings in spoken and written forums such as debates, conferences, newspaper articles, and television programs during most of the 1980s can be explained by two factors: since textbooks are the most frequently used educational material their content was immediately understood; and the stereotypes found in these materials were cartoonish. At the same time this depiction of the stereotypes created negative reactions in female readers since the content was both close to women's experiences and remote, which generated indignation among the women. Official organizations, such as the Council on Women's Status, the buyers and distributors of books (the then-FLE, and its successor FDE, in São Paulo), and the State and Municipal Departments of Education, organized seminars and materials on this issue. Despite these activities, there were some gaps in the studies that were not addressed: the use teachers and students made of these books; the difficulties encountered in creating alternative material and what that means in Brazilian society; how these sexist products were used; and the implications of the fact that in Brazil, the state is now the largest buyer of textbooks.

The topic appears to have exhausted itself in the 1990s, given the lack of studies with a diachronic perspective that tracked how forms of gender discrimination changed in teaching materials, despite the importance, in the current administration, of the National Textbook Program.

Although we lack updated research studies on sexism in textbooks, Escanfella and other collaborators[39] did a longitudinal study analyzing changes in children and young adult books using the lenses of gender, race, and age relations. The purpose of their study was to show what has remained constant and what has changed in the construction of the characters in contemporary Brazilian children's books (1975–1995) compared to an earlier study which analyzed 168 books published between 1950 and 1975. In both studies, the analysis focused on gender, race, and age discrimination as reflected in the characters. Using the theoretical model proposed by John B. Thompson[40] to analyze ideological constructs, the study was divided into two phases: first, it examined the attributes of 1,800 characters in 60 stories contained in 41 Brazilian children's books first published between 1975 and 1995; and second, it compared certain selected indicators with those of the earlier study.

On the issue of gender discrimination, Escanfella noted that in general, both studies were very similar, with no change over time found in the general profile of male and female representations. Male characters are represented more frequently, and continue to assume a relatively important position on the fictional and social level; female characters assume a relatively important position in family relations. However, a decrease in the intensity of gender discrimination was observed.

Beisiegel's analysis[41] of MEC's textbook evaluations in light of current policies of the PNLD, affirms the seriousness of the work performed by the teams of specialists. He reported that a small number of books were discarded because of their prejudices and stereotypes, and concluded that, with few exceptions, the expert opinion of the PNLD on ethnic and gender issues used the same starting point as the analysis done of school textbooks in the 1950s and 1960s. Examples of more subtle discrimination were not captured in MEC's current evaluation because their methodology had not been updated to reflect current stereotypes and forms of discrimination.

It should also be noted that the research studies and interventions regarding textbook stereotypes place an emphasis on elementary school education, omitting books intended for secondary and undergraduate levels. No studies have been performed about the gender dynamics within the rich and extensive editorial market that was created for, and around, the expanded educational system, or the emphasis placed on the textbook as an instrument to improve education, both of which are crucial questions at this time.

Early Childhood Care and Education

From the earliest stage of the feminist and women's movements, demands were made for early childhood care and education (ECCE) to supplement maternal care, especially through child care centers.

The decade of the 1990s can be divided into two periods: the first lasted until 1996, when a strong movement could be seen in Brazilian society that expanded preschool education; the second period corresponded to the recent refocusing of national priorities toward elementary school education, which slowed down the growth of preschools.

At the legislative level, the Constitution of 1988 and the 1996 National Education Law are both noteworthy. It was the Constitution of 1988 that, for the first time in Brazilian history, recognized the right of a small child to be educated outside the family in child care centers and preschools. The National Education Law also included preschools in the educational system, another first in Brazilian history. As a result, child care centers and preschools became regulated by the educational administration; their objectives were defined, curricular guidelines were established, and minimum training requirements for staff were specified. The allotment to be set aside in municipal budgets and the goals to be achieved according to the National Plan of Education were specified as well.

Despite the serious shortcomings in the educational statistics at this level,[42] a significant growth in matriculations is observed for the years 1986–2000, increasing from 4,177,302 in 1986 to 6,012,240 in 2000 (a 43.9 percent jump). The importance of this spectacular increase must be qualified, however, to the extent that there is still a large contingent of preschool teachers who lack the required educational credentials (see table 5.8).

Table 5.8 Teachers in early childhood who have less than a secondary school education—Brazil, 1986 and 2000

Preschool and literacy classes—1986		Daycare, preschool, and literacy classes—2000	
No.	%	No.	%
48,269	27.5	320,841	13.0

Source: Educational Census 1986 (cited in Rosemberg, 1999, p. 22) and Educational Census 2000.

It has also been observed that the priority given to elementary school education under the Cardoso administration slowed the pace of the expansion that took place in the 1980s. Additionally, evaluations of annual cost per student have shown that the large gap between the cost of educating a student in early childhood education (US $820 per year) and in higher education (US $10,791 per year) has not narrowed. Such a difference in cost per student supports, creates, and reproduces structural inequalities of gender and age.

CONCLUSION

This chapter should be accompanied and supported by a theoretical synthesis that attempts to link feminist and educational theories. This, however, was not my intent. My aim was to point out the dominant as well as ambiguous trends in the empirical data. Over the course of 2001, I produced a series of articles,[43] in which I attempted to demonstrate and interpret the gaps in the production of knowledge within the fields of Education and Women's/Feminist/Gender Studies. In this preliminary exercise, I came up against a profound ignorance of the situation of men and women within the educational system and of the male and female dynamics therein. I also encountered misunderstandings that are unfortunately shared by progressives and conservatives alike in organizations such as the women's movement and the World Bank.

The fear of encountering ambiguities and contradictions has given rise to interpretations about men's and women's education that essentialize their differences. In my opinion, these cloud our interpretation of the complex paths of structural and ideological gender domination that is intrinsic to contemporary societies. In this chapter, I believe I have taken an important step by widening the agenda of research studies beyond the space of the school. There, perhaps, we will be able to perceive new processes of gender, class, race, and age domination.

GLOSSARY OF ABBREVIATIONS

CNDM: Conselho Nacional dos Direitos da Mulher
DAES: Divisão de Avaliação do Ensino Superior
EFA: Education For All
ENC: Exame Nacional de Cursos
FUNDEF: Fundo de Manutençãoe Desenvolvimento do Ensino
IBGE: Instituto Brasileiro de Geografia e Estatística
INEP: Instituto Nacional de Estatística e Pesquisas Educacionais

IPEA: Instituto de Pesquisas Econômicas Aplicadas
MEC: Ministério da Educação e do Desporto
PCN: Parâmetros Curriculares Nacionais
PNLD: Programa Nacional do Livro Didático
PNUD: Programa das Nações Unidas para Desenvolvimento
RAIS: Relações Anuais de Informações Sociais
SAEB: Sistema de Avaliação do Ensino Básico

Notes

1. Student's ages often do not correspond to a standard grade level.
2. Lei de Diretrizes e Bases da Educação Nacional.
3. Fundo de Manutenção e Desenvolvimento do Ensino Fundamental e de Valorização do Magistério.
4. The expression "Washington Consensus" was introduced by the economist John Williams to describe the vision that corresponds to the position taken, at the end of the Cold War, by the government in Washington, research centers, and the offices of international economy that have their central offices in that city. The terms of this consensus are that Washington should direct the architecture of the new world economy in accordance with the forces of the free market. (Manuel Campeses, Jr., "O futuro do consumo de Wasington," www.esg.br/publicacoes Artigos, 2002: 1–10).
5. Brasil, MEC/INEP, EFA 2000, *Educação para todos: avaliação do ano 2000*, Informe nacional, Brasilia, 2000: 15–16.
6. In Brazil, the women's movement and the feminist movement have different ideological agendas and are treated as separate movements.
7. I use the term "race" in the sociological sense to refer to the attributes ascribed to individuals for the purpose of classifying them sociologically. The term is therefore not used in a biological sense.
8. Pesquisa Nacionais para Amostra de Domicílios.
9. Ministério da Educação e do Desporto.
10. Instituto Nacional de Estatística e Pesquisas Educacionais.
11. Relações Anuais de Informações Sociais.
12. A recent initiative of the National Council on Women's Rights (*Conselho Nacional dos Direitos da Mulher*/CNDM) should be praised because it is an exception to the usual precarious nature of educational data that accounts for sex. The project, *Indicadores de Gênero*, is available at the CNDM website (www.mj.gov.br/sedh/cndm). It includes an excellent overview of statistical data on education.
13. Advancement refers to the pace and the appropriateness of grade level and age. Many Brazilian children are behind for their age because they have had to repeat levels.
14. These branches include new trajectories and innovations such as the accelerated classes (two grade levels in one year, for example).

15. 1985 and 1999 PNADs.
16. The differences were on the order of 8 percentage points.
17. Alceu R. Ferrari, "Analfabetismo no Brasil," *Cadernos de Pesquisa*, 52 (1985): 35–49.
18. Data with a breakdown by sex is unavailable about enrollments by young adults and adults for education courses (Census on Young Adult and Adult Education). The results of the 2000 Census are awaited in order to study the incidence of the return to illiteracy of older adults, a little-studied phenomenon in Brazil.
19. 1999 PNAD.
20. With the exception of ages five to seven.
21. 1999 PNAD.
22. 1999 PNAD.
23. Brazil, MEC/INEP, EFA, 2000, *Educação para todos*.
24. *Ensino supletivo* replaces regular school for teenagers and adults who have not concluded their studies by the expected age. It also provides refresher or review courses for those who have followed, either totally or in part, the regular school program. These classes are held in schools or are imparted by radio, television, or correspondence, with exams administered to students at the elementary to university levels, as the case may be.
25. 1999 PNAD.
26. 1929, 1944, and 1993.
27. Erika Lourenço and Mônica Jinzenji, "Ideais de crianças mineiras no século XX: mudanças e continuidades," *Psicologia: teoria e pesquisa*, XV, 16, 1 (Jan.-April, 2000): 45.
28. Relações Anuais de Informações Sociais.
29. Wanderley Codo, *Educação: carinho e trabalho*, Brasília/Petrópolis: UNB/Confederação Nacional dos Trabalhadores em Educação, Vozes, 2002.
30. Fundo de Manutenção e Desenvolvimento do Ensino Fundamental e de Valorização do Magistério.
31. Brazil, MEC/INEP, *Censo do Professor1997*, Brasília, INEP, 1998.
32. Conselho Nacional dos Direitos da Mulher.
33. Brazil, Presidência da República, *Programa Nacional de Direitos Humanos*. Brasília, 1996, 17.
34. Parâmetros Curriculares Nacionais.
35. Brazil, MEC, *Parâmetros Curriculares Nacionais Temas Transversais 3° e 4° ciclos do Ensino Fundamental*, Brasília, MEC/SEF, 1998.
36. Programa Nacional do Livro Didático.
37. Celso R. Beisiegel, "Uma Cultura para a Democracia," São Paulo, FEUSP, mimeo, undated.
38. Esmeralda Negrão and Tina Amado, *A Imagem da Mulher no Livro Didático*, Textos FCC, 2, São Paulo, FCC, 1989.
39. Célia Escanfella et al., *Literatura Infantil e Ideologia*, São Paulo: PUC-SP, mimeo, 2001.

40. John B. Thompson, *Ideologia e Cultura de Massas* (Petrópolis: Vozes, 1995).
41. Celso R. Beisiegel, "Uma Cultura para a Democracia," p. 29.
42. The difficulties stem from two factors: the existence of a "clandestine" network of child care centers and preschools that are not listed in the registers of the Secretary of Education; and the fact that it was only in 1995 that children aged 0–4 years were included in educational measures on the questionnaires used by the IBGE (Rosemberg, Fúlvia, "Expansão da educação infantil e processos de exclusão," *Cadernos de Pesquisa*, 107 (1999): 7–41).
43. Rosemberg, Fúlvia, "Políticas Educacionais e Gênero: Um balanço dos anos," *Cadernos Pagu*, 16 (2001): 151–198. Rosemberg, Fúlvia, "Educação formal, mulher e gênero no Brasil contemporâneo," *Revista Estudos Feministas*, 16 (2001): 151–198; Rosemberg, Fúlvia, "Caminhos cruzados: educação e gênero na produção acadêmica" *Educação e Pesquisa*, 27, 1 (2001): 47–68.

CHAPTER 6

The Feminization of the Teaching Profession in Belgium in the Nineteenth and Twentieth Centuries

Marc Depaepe, Hilde Lauwers,
and Frank Simon

In the prologue to John L. Rury's book, *Education and Women's Work. Female Schooling and the Division of Labor in Urban America, 1870–1930,* published in 1991, Barbara Finkelstein writes: "The history of education is not rich in studies that combine an effusion of quantitative data, with equal portions of narrative elegance, biographical perspective, and attention to intellectual and spiritual as well as material dimensions. *Education and Women's Work* is such a book."[1] This affirmation is somewhat frustrating for the Belgian investigation into this topic.

More or less at the same time, in our book about the social position of Belgian men and women teachers in the twentieth century,[2] we wrote that the investigation into the feminization of teaching in our country was still in its initial stage, and that the concept of "feminization" not only meant that the feminine element had become dominant in the teaching population in quantitative terms (in other respects, an ahistorical observation). We observed that the fluctuations in the ratio of men and women within a profession should never be considered independently from the circumstances that determine social development, such as economic conditions and the labor market, social and cultural characteristics, social origins, religion, cultural norms and values, dominant social models, inequality between the sexes, professionalization, degree of education, the length, value and status of the teacher training, urbanization, demographics, wars, legislation, and so on.

Ten years later, in 2001, we could affirm that, within the framework of the historiography about changing male/female ratios in Belgium, some attention had been paid to education, in the same way that some interest had been shown toward the differentiation by gender in the field of pedagogical historiography. Unfortunately, these two fields bore few results, and one searched in vain for research that had as its focal point the development of education with a gender thematization.[3] Thus we concluded that it was not possible to present a general panorama of the feminization of the teaching corps in Belgium at that time. We lacked studies that would permit us to form a reasonably complete image. In those ten years we had made little progress.

It was at that point that the "First International Congress on the Processes of Feminization of Teaching" (San Luis Potosí, Mexico, February 21–23), 2001 was held, at a very opportune moment in the Belgian context, as well as internationally. In collaboration with the network established there, we now hope to be able to develop interpretive models in the future, and also to elaborate a comparative historiography that is founded on a more consistent methodology. At present, however, we should be satisfied with the very limited data we have. The only source that we can truly draw on are a series of relatively homogeneous statistics that have been published on preschool, primary, and (to a lesser degree) secondary school in Belgium, since 1830.[4] These constitute the nucleus of our chapter, which as a result— and against our will—has a merely descriptive character. However, as a first step toward a more detailed investigation, we will work with the following restricted definition of feminization: the feminization of teaching as measured by the increased percentage of women in the teaching population. All too many times we have had to specify that a large part of the research into feminization is performed in a vacuum, that is, without the necessary empirical evidence.

In this chapter, apart from some preliminary considerations with respect to women and the labor market, mostly statistical data will be presented on women and men teachers (profession and education). Then, with a view toward a strictly historical interpretation, we will also focus our attention on the position of the woman professor or teacher in the entire teaching corps. We will briefly examine unequal salaries, professionalization (supplementary training, participation in unions, cultural production, etc.) and some recent phenomenon.

However, to evoke the mentality in which the described quantitative evolution has occurred as historical processes, we will begin with several eloquent declarations about the presence, or in this case, absence of women teachers in Belgian education. The first two quotes come from our source of "life interviews" with individuals from those times. We give the words of two former primary school teachers, both men.[5] The third quote expresses the opinion of a Belgian feminist.

In the first case, Julien Van Overbeke, born in 1899 and the headteacher of a municipal primary school from 1950 to 1960, recalls that the women teachers in his school were a threatening presence for the men:

> We had good relations with them, but from a distance, because women are dangerous. In our school, there were twenty teachers, twelve of whom were women. And you know what women are like, young women. And we weren't old men, either. They wanted to put you in a "compromising situation" so that they wouldn't have any-thing to do. Yes! Yes! The eternal Eve. I kept my distance, because it was dangerous.

The second quote is from Rogier Carmen, born in 1911. He worked in education from 1932 to 1961 and was the provincial secretary of the Christian syndicate. He thought that women—he was talking about young women—should not give classes to young boys:

> I think it's a shame that these days there are also women in schools for boys. I think that young women have a difficult time with boys because you have to be firm with some of these brats. They're kids who at thir-teen or fourteen want to be the center of attention, and they turn the class upside down. They challenge the teacher, and say disrespectful things. It's terrible for whomever is in front of the class.

The third quote is from a recent publication by Renée van Mechelen, who during the last quarter of the twentieth century defended women's liberation:

> A male bastion, that's how the entire educational sector could be described. However much it's been feminized, the men are at the top. As I was following the television coverage about the problems in our educational system, I was struck by the fact that at demonstrations one sees many women, and at the bargaining table practically none. Now, as educational system is being rationalized there will be a struggle for the principal's position, but we can rest assured that these positions will be given to men. In the long run, one should oppose the idea that a girls' school and a boys' school enter into a collaborative relationship, because the odds are strong that the principal of the boys' school will be named principal.[6]

With these testimonies we intend to underline how insecure the position of women in education was in the past. She was first judged by her condition as a woman before her condition as a teacher. Although not openly, this sexism has, without a doubt defined the position of women teachers for a long time. The degree to which this discrimination continues to have an influence today has been shown by some studies, and later refined by others. In each case, new biographical studies[7] have shed more light on the choices made by women. However, these studies are still so limited in number that it is impossible to draw any general conclusions, at least in the case of Belgium.

Despite the opinions quoted above, the feminization of the teaching profession in Belgium at the beginning of the twenty-first century is stronger than ever. At the present in Flanders (the Dutch speaking part of Belgium), in more than four of five cases in primary school education, a woman heads the class. In secondary education there are three women for every five men teachers. In these two levels of education, feminization is growing. In schools of higher education that are not universities, we find hardly two women for every five teachers. And at the university level, the ratio doesn't even reach one woman for every five faculty members.

How Should We Interpret This Data?

From international research studies we see that the progressive feminization of teaching, as measured by the increased percentage of women in the teaching population, is a phenomenon that begins when the nineteenth century is well underway. These trends seem to be pointed in the direction of stability and uniformity. However, this does not eliminate the fact that the fluctuations in the ratio of men and women in a professional group can never be considered in isolation from its historical context. There are no unequivocal explanations. Gender inequality in the labor market seems significant to us.[8] For example, for a long time women teachers received lower salaries, which made them more attractive to employers. Men, on the other hand, benefited from a wider spectrum of alternatives when it came to choosing a profession, which may have caused them to show less interest in teaching. But is this sufficient to explain the great variations that exist in the degree of feminization in different countries?

The historical context has undergone such changes over the last hundred years that the relative influence of diverse factors in this period most likely has not remained the same. In the international research, the process of feminization has also been described as an extremely complex phenomenon. In Europe as well as in the United States, the teaching workforce is strongly feminized. James Albisetti showed in a comparative study that this uniformity, however concealed, great differences. Toward the end of the nineteenth century, Russia and Italy also had approximately the same percentage of women—70 percent, working in primary school education as in the United States and England. Germany, which was undergoing rapid industrialization, and Denmark and Holland, had a significantly lower percentage: approximately 30 percent. Meanwhile, countries such as France and Belgium were at 50 percent. Albisetti reached the conclusion that in distinct countries with significant differences in the level and general context of feminization, similar explanations were found. Economic conditions, legislation, religion, job termination upon marriage, cultural traditions, gender ideologies, urbanization and war were all cited as possible explanations.[9] The only question is whether in this study, Albisetti addresses the essence of the international process of feminization. In our opinion, the interaction between the quantitative and qualitative factors of feminization are, to a large extent, related to the dynamic inherent to pedagogical activity in education itself, as we shall see further on.

WOMEN IN THE LABOR MARKET

As it has been noted, the differences in employment opportunities for men and women in teaching cannot be understood without taking a look at the evolution of the labor market on the whole.[10] Above all, possibilities for women on the labor market can vary considerably from one period to another. The question is: why do women decide on education at a given moment and, by contrast, why do men, to a lesser extent it seems, choose a career in education?

When Belgium became independent in 1830, a liberal constitution was established that guaranteed the independence of the judiciary and freedom of religion, association, the press, and education. Nevertheless, the progressivism of that era did not do a thing to improve the position of women. The spirit of the *Code Napoléon*, which equated women with minors and declared them irresponsible, was incorporated into Belgian law. Moreover, in all types of writings that date from this period, fairly reactionary ideas about women appeared, ideas that were voiced not in small part by women themselves.[11] And despite the fact that the female half of the population played a market after the mid-1960s paralleled the decrease in the so-called civilization offensive in the middle classes, for whom the ideology of the woman in the home constituted an important element.[18]

From the research we will see later that women are not equally represented in the labor market, neither in the nineteenth century nor in the twentieth, a situation that, all things considered, persists until today. The labor market in its totality was characterized by a vertical and horizontal segregation by gender. Women did not have proportional representation across the sectors, nor were they represented proportionally in different jobs within the varied enterprises. In the nineteenth century, with respect to the manufacturing industry and work done at home, we find women in the textile, clothing, and lace industries above all. In the commercial sector, we find women in food products and in the restaurant and food services sector in particular. One could say that the position of women at home is replicated in the labor market.[19]

In the first half of the twentieth century, we still find a great proportion of women in a limited number of professions. Men seem to have an unlimited choice of profession, while the options open to women are much more restricted: more than 90 percent of working women occupy only eight professions. Typical professions for women are concentrated in the helping sector (and within the helping professions, in the health and education sectors) and in the professions related to domestic work (personnel for housework and kitchen personnel). One finds these professions reflected in the list of the skills that are expected of a housewife.

And just as in the nineteenth century, we still find a strong vertical segregation today. Women are over-represented in the lowest positions and are scarcely found in the higher echelons.[20] Apparently the mentality of the previous century or centuries strongly persists. In 1943, Maria Schouwenaars, a distinguished Catholic feminist and school inspector, published the book, "*Wat zal ons meisje worden?*" (What Will Our Girl Become?), in which she very clearly summarized the general mentality. Her primary answer to the question in the book's title was: "before all else, an energetic housewife!" According to Schouwenaars, the best form of education was still vocational education, which would not only prepare a girl for a female profession, but would also contribute to her happiness and *joie de vivre* by offering her a chance to develop her "feminine psyche,"[21]—a cultural historical factor that we will use as a conceptual tool in our conclusions.

Statistical Data on Women and
Men Teachers

In discussing the feminization of teachers, it makes sense to examine the data first. For secondary and higher education we have very fragmentary statistics for both the nineteenth as well as the twentieth century. Separate statistics for primary schools and teacher training colleges were published in the past. Fortunately for our research, a breakdown by gender is one of the few categories that has enjoyed a secure continuity in the statistics that are published every year.

Beginning with the second half of the nineteenth century, we can observe a gradual increase in the number of women teachers (see figures 6.2 and 6.3). The sudden reversal of this trend in the 1880s was a result of the so-called School Conflict. During the first School Conflict (1879–1884), the administration, under a liberal government, no longer published data on all Catholic schools. And the same happened during the period between 1884–1894 (under a Catholic government), meaning that the available data must be interpreted with caution. Bear in mind that, due to the inconsistent publication of the statistical series, the percentages mentioned here (relative figures) do not always refer to the total population of teachers (absolute figures). However, our analysis of the trend is representative of teachers as a whole.

The turning point at which, for the first time, more women teachers are registered, occurs almost at the turn of the century, specifically

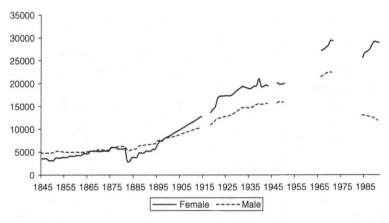

Figure 6.2 Number of teachers in primary school education by sex (1845–1992)

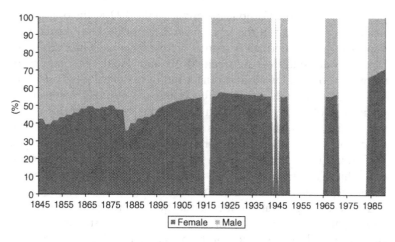

Figure 6.3 Percentage of women teachers (1845–1992)

in 1898. However, the relatively strong increase in the number of women teachers was not a consistent phenomenon during the first half of the twentieth century. After 1923 the rate of feminization stabilized, while the absolute number of women teachers kept growing. Was the increase in the number of women teachers in the immediate aftermath of World War I caused by a lack of men teachers? Probably not. In peacetime, men teachers were exempted from military service and during the war they were placed in the rearguard.[22] As a result, the number of men teachers killed in battle must have been very limited. Was this increment primarily due to the establishment of compulsory education in 1914, thus increasing the number of classrooms? We do not have any statistics regarding the number of classrooms. In any case, the number of schools, and more specifically the number of schools for girls, did not grow at a spectacular rate after World War I.

The relatively sudden stabilization of the number of women teachers after 1923 can be explained by the fact that, in that same year, Belgian bishops announced that they intended to maintain the celibacy requirement for women teachers in Catholic schools. Were many women alarmed at having to make such a drastic choice at a young age? In the 1920s, moreover, there was renewed resistance to hiring women teachers in boys' schools.[23] Furthermore, the 1930 papal encyclical *Casti Connubii*, which again emphasized the blessings of procreation, conjugal happiness, and the separate roles of men and women, must have had an influence not so much on the Catholic population as on the Catholic hierarchy. In the 1920s, the overall participation of women in the labor market increases. Did women perhaps have better alternatives than a job in education? Once again, we lack relevant statistics. Between 1920 and 1930 the number of women who found work in business and in industrial jobs rose more than 80,000 in each category. Regarding the strong growth in the percentage of women teachers after 1970, no studies exist that correlate this increase with the labor market overall, with the changes in the education of men and women, or with social changes. Perhaps, then, the feminization of teaching is a recent phenomenon? Whatever the case, this phenomenon has only recently been detected and problematized in the field of education.

However, a general outline of the quantitative evolution of the female participation among primary school teaching staff is not sufficient to provide a clear vision of the phenomenon. A more diversified approach is required. And this in no way relieves us from a tedious analysis. In the case of Belgium, teaching is firmly structured on ideological fracture lines. A relationship can be established between this structure and the characteristics of feminization in teaching.

First, since the beginning of the twentieth century, we find that the majority of women teachers worked in Catholic education (in figure 6.4, Catholic education is shown by striped horizontal bars). This in part is explained by the number of women in religious orders, which we will return to later. Catholic schools could easily rely on nuns, while lay women teachers had access, at the most, to subordinate positions.[24] Yet even after the 1970s, when nuns decreased in number until becoming an insignificant minority, this sector kept attracting many women, The process of laicization (the decrease in the number of teachers belonging to religious orders) has been hardly studied. An international literature on this theme, comparable to the one that exists on feminization, has yet to be written.

If we take as our point of departure the ratio of nuns to lay women teachers, we can observe that it was at its peak in 1920 (see figure 6.5). One out of every two women teachers in primary education was a nun, an extremely high ratio. Likewise, the absolute numbers show a strong increase around the turn of the century in the number of nuns. In part, this increase can be attributed to the immigration of French nuns as a result of the 1904 French law promulgated by Interior Minister Emile Combes that prohibited nuns from working in education in their native country.[25] In several caricatures of the era, the arrival of the nuns from France was represented as a "black invasion."[26]

This spectacular increase in teaching nuns has been criticized more than once. In 1905, 35 percent of the nuns working in denominational education did not have a teaching degree (compared to 2 percent among lay women teachers), and they were strongly reproached for the ease with which they took teaching jobs away from lay staff. The situation was most likely the same in preschool and early childhood education, areas which were totally feminized and in which a large number of nuns without degrees had cornered the market.[27] The increased number of nuns around the turn of the century was not only noticeable in teaching. Religious vocations grew in general. In Belgium, in 1866, there were almost 14,000 nuns; by the turn of the century, their ranks had doubled, and by 1910, the country had more than 47,000 nuns.[28] Normally, in nineteenth-century women's studies, the focus is placed on the growing importance of home life and motherhood, while the growing popularity of religious life has remained outside the center of attention.[29] As Josephine May has recently indicated in relation to the Australian situation in the 1930s to the 1950s, the powerful femininity of the female teachers consisted in large part in the image of the "secular" nun.[30] Whatever the case, in the twentieth century the rate of more than 50 percent was never equaled after 1921. Right before World War II, the ratio of nuns in the teaching population was only a third of the total. In absolute figures, their numbers remained above 5,000. The noted decrease in the percentage of teaching nuns can be almost entirely attributed to the fact that they could no longer keep up with the rate of growth of lay teachers, and not to a dramatic decrease in their numbers.

It is not until more or less after World War II that one can speak of a true decline. Between 1943 and 1954 the population of nuns in primary education was reduced by almost 15 percent. With respect to developments after 1954, complete studies have not yet been done. However, from the research one can observe that starting in the 1960s the ratio of nuns falls exponentially, which evidently coincides with the steep decline in the number of vocations and the growing renunciation of religious vows.

From this data one sees that the massive presence of nuns in primary education initially had a favorable influence on the degree of feminization. Given that the drastic decrease in the number of nuns in the 1960s and 1970s did not give rise to a declining feminization, we must admit that the so-called mechanism of "synthetic turnover"—in which lay women took the place of nuns—was also a factor in Belgium. The available statistical data, however, is not sufficiently refined to demonstrate this in mathematical detail.

The Feminization of Teacher Training Schools Prior to the Feminization of the Teaching Field

The feminization of teacher training followed a very clear process that may have something to do with its shorter length and, therefore, with its greater sensitivity to changing trends in this population. The growing number of students in teacher training schools can be interpreted, in the first place, as a result of the increased number of training schools. When in 1842 the first law concerning primary school education took effect, teacher training was not mentioned. Toward the end of the 1840s, a need for certified women teachers was felt within the growing field of girls' education. In the nineteenth century, policies were oriented toward creating separate schools for boys and girls. Women, consequently, were preferred for jobs in girls' schools.

Already, by 1860, more than half of all educational institutions were for girls, and by the mid-1870s there was a ratio of six girls' schools for every four boys' schools. Feminization, which arrived with the increase in teacher training schools for women, was further consolidated during the interwar period. The feminization of education that occurred among the trainees was indeed spectacular. In 1849, girls still made up only 11 percent of students enrolled in teacher training schools, but in the following years the figures would change drastically. In scarcely 25 years, gains for women would be such that by 1874 they would make up more than half of the total number of students in training schools. And likewise, in the following ten years, the number of women students in these schools continued to grow. Between 1885 and 1900 the proportion of women was already at 55 percent and 60 percent. The fact that the percentage of women students grew to such a degree, without a corresponding growth in job opportunities, indicates that enrollment in a teacher training school did not automatically guarantee that one would step immediately into the profession. And both these factors point to the fact that the teacher training school was the only escape from waiting for Prince Charming to arrive. For young women, continuing their studies and going to teacher training school were one and the same thing.[31]

It is clear from figure 6.6 that by the end of the nineteenth century, the number of women who were awarded degrees in teacher training schools exceeded the number of men. Between 1911 and 1921, registrations doubled in teacher training schools for girls. A change in the class background of the students provides us with an

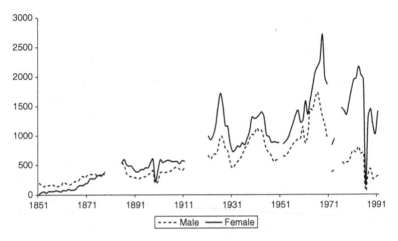

Figure 6.6 Total number of degrees from teacher training schools by sex (1851–1992)

explanation for this phenomenon. After World War I, training schools also began to attract girls from less-privileged classes. Likewise, the lack of male candidates for marriage after World War I may have forced girls to guarantee their own income. A career in education secured their future. But when in 1923 and 1926, a series of changes in the teacher training schools resulted in a longer and more expensive education, the number of girls dropped. The girls from the lowest classes were the ones who had to give up this more expensive education. After 1936, the teaching of education to girls was again on the rise.

The fluctuations corresponding to men and women with education degrees parallel each other closely until the end of the 1960s. At the start of the 1970s, there is a significant downturn in the number of male and female graduates with education degrees. In 1970, education became a two-year course of study upon completion of secondary school. With this, a change that had already begun was officially confirmed. It is unknown to what degree this may have influenced the number of matriculations in teacher training schools. The number of women with degrees returns to its previous levels after a few years; however, the number of men with education degrees remains very low from that point on. As a result, in 1998 more than eight out of every ten education graduates is a woman.[32]

The feminization of teacher training thus began before the feminization of the profession. Over the last twenty years this tendency has become clearly accentuated. This evolution parallels the expansion of education that dates from the 1960s, which brought a greater degree of schooling to all children. While in the 1960s the degree of schooling for girls still lagged far behind boys, over the course of the 1970s, girls completely caught up with boys at the secondary school level.[33] The fact that, at the beginning of the 1970s, girls conformed to the classic feminine role when choosing their field of study can be seen in a survey performed in those years. In principle, their parents had nothing against their continued studies. However, they considered that their daughters' primary responsibility was to their future family.[34] A career in teaching, therefore, could be a partial solution, as long as work and family were compatible, as seen from the point of view related to content, as well as practice.

Another possible explanation for the more intensive and earlier feminization in teacher training in comparison to the teaching profession itself is that the average length of study to become a teacher is between three and four years, while the years in which the profession itself is exercised is almost ten times as great.[35] At this time we do not have exact figures regarding the duration of a woman's professional career, but this does not prevent a study from being done which examines its length for married or, as the case may be, single women. Moreover, until the end of the 1950s or the start of the 1960s, the majority of women teachers who worked in Catholic schools were dismissed as soon as they married, as ordered by the bishopric. In order for Catholic schools to find sufficient women candidates to replace their married colleagues, the supply of women teachers had to be greater than the supply of men teachers.

However, just as we did for the labor market, we must now perform a comparison within the female student population as a whole, in relation to the professional choices open to women. Studying the feminization of education in Germany,[36] Peter Lundgreen reaches the conclusion that in 1931, 24 percent of all male students chose a career in teaching, compared to 53 percent of all women students. Thirty-five years later, these figures are 20 percent and 55 percent, respectively. And another thirty years later, student preference for this field of study drops to 4 percent and 14 percent, respectively. In this way, at the moment when the number of women students surpasses the number of male students, the preference for teaching is at its lowest point in history. At the end of the twentieth century, the majority of women in Germany apparently chose another profession. Seen from the perspective of the overall female student population, teaching can no longer be called a female profession with total certainty.

We still do not have any comparative studies for Belgium. Recent figures, however, support the conclusion that the situation is different, at least in the case of Flanders. In 1999, almost 20 percent of girls who graduated from high school chose to take teacher training courses to prepare them for the teaching profession, while for boys this percentage hardly reached 9 percent. At the university level, 19 percent of women students took courses which would give them access to a career as a teacher in secondary schools or training institutes, compared to 9 percent of male students.[37] However among Belgians 22 to 30 years old, only 2 percent chose a teaching career.[38]

Position of the Woman Teacher within the Teaching Corps

The situation of "Miss" or "Sister" may seem positive in absolute terms, but within the teaching corps women did not have the most advantageous position. After all, salaries for women teachers during the nineteenth and a large part of the twentieth centuries, were inferior to the salary received by their male colleagues. In 1895, the maximum base salary for a woman teacher was almost 7 percent less than the maximum salary of a male teacher. In 1914, this difference ascended to 20 percent, to fall once again to hardly 2 percent in 1935.[39]

The salary disparity between men and women teachers was no exception to the rule. Women's work received less compensation in other professions as well. Besides the logical cost savings a lower salary brings, a woman's salary fulfilled, in the eyes of her contemporaries, a different role than a man's salary. It was seen above all as supplementing a family's income.[40]

From studies on work in factories in the nineteenth century, one observes that men's work was more quickly deemed skilled work, even when in truth it was not. The skills that male workers were required to have were called "specialized knowledge," and as women did not possess such knowledge, a higher salary for men was justified. The skills held by those who performed "women's work" were not acquired, but innate. These were seen as an extension of skills women employed in the home, for which they received no remuneration. In education we find a similar line of reasoning. Along with a great variation in teachers' salaries depending on, among other factors, the region (city/countryside), the board of administrators, marital status, job function, and title, a distinction was also made on the basis of gender.

Before 1914, salaries for men and women teachers were so low that approximately 40 percent of men teachers gave evening classes to adults, compared to 16 percent of women teachers. In cities, there were more opportunities to round out one's salary, not only with evening studies, but also with private classes. It is not known to what extent women teachers added to their income in this way.[41]

Then in 1920 salaries for male and female primary schoolteachers were made equal by law. In Belgium women teachers were the first women to receive an equal salary for equal work![42] However, teachers belonging to religious communities or unmarried teachers were still paid a lower salary. This naturally had a greater impact on women, as the majority of teachers belonging to religious orders were women, and women teachers in Catholic schools had to remain single. With the economic crisis of the 1930s and the resulting cost-cutting measures adopted by the administration, women teachers lost their right to an equal salary. They were not the only women to suffer the consequences of governmental budget cuts: in 1934 a hiring freeze was ordered for women in government jobs, and women already working for the government were asked to take a 25 percent pay cut. After World War II, these salary differences were eliminated, with the exception of nuns who continued to earn half the regular salary.[43]

The feminization of the teaching profession is cited regularly to explain its low status. Nevertheless, women's contributions to the professionalization of teaching, both in the production of knowledge and values as well as in the definition of professional attitudes, cannot continue to be ignored. In the end, the teaching profession was one of the first professional labor activities that, at least from the turn of the twentieth century on, swelled its ranks with the female population. Research studies, both national and international, on professionalization trends in the nineteenth century normally do not take into account the female factor without explaining women's inferior status, as mentioned earlier.[44]

Professionalization, academic skills, ethical dimension, and deontology are terms that have been used by different players in the education field. Unions and other teacher organizations utilize them in order to demand a greater esteem for their profession and, as a result, a higher salary. But the administration also has used these terms on several occasions to disguise their control over teachers, by emphasizing teacher autonomy.[45] Recently, German researchers have advocated the use of the term "Verberuflichung" (meaning "professional skills as such"), given that *"Professionalisierung"* (derived from "profession") seemed too ahistorical.[46] Nevertheless, the term professionalization is still frequently utilized in historical research studies. And although the content of this term may change depending on the social context,[47] (and as we mentioned above, depending on which player uses it) the term "professionalization" has accompanied the teaching profession ever since the mid-nineteenth century. Most of the time, what is meant is that the teacher acquires a specified measure of control over the exercise of his or her profession.[48] Contributing factors are his or her educational level, in-service training, teachers' associations, writing school textbooks, and so on. Unfortunately, the picture that we sketch here of the situation in Belgium is again incomplete and based on fragmentary evidence.

Whatever the case, professionalization was advanced when the teaching degree was made mandatory. With this requirement also came a legal framework regarding access to jobs, conditions and procedures, and the kinds of tests and examinations required to obtain a teaching position.[49] Here the state played a fundamental role as the employer of these graduates and as supervisor and guide to new developments that arose in the profession.[50] The number of teachers with degrees working in education between 1860 and 1929 exhibit a significant difference between men and women teachers.

Figures 6.7 and 6.8 show that between 1860 and 1929, the total number of women teachers without a teaching degree was much greater than that for men teachers. And it is women in religious orders who by far taught classes without a degree, despite the strong increase in the number of students taking courses in education. Apparently, the majority of Catholic schools for girls preferred women teachers with

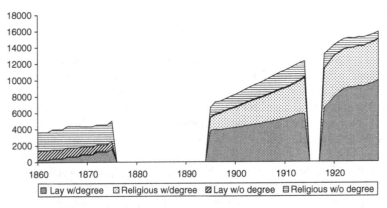

Figure 6.7 Total number of women teachers with degrees (lay/religious) (1860–1929)

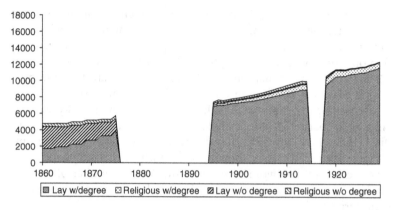

Figure 6.8 Total number of men teachers with degrees (lay/religious) (1860–1929)

no degree, a "sister," who would be much cheaper than a lay teacher who had a degree. In the end, she would be more economical and ideologically more trustworthy.[51]

We also find homogeneous tendencies in other aspects of professionalization.[52] Take for instance the opportunities for in-service training. The idea that women would have little interest in it or that they wouldn't have free time because of family obligations, must be placed in context. In 1938, in the province of Limburgh, 120 teachers enrolled for a course in the Higher Institute of Educational Sciences. However, the number of students exceeded the anticipated number and went beyond the capacity of the Institute. Following the Inspector's recommendations, the Institute ordered women teachers to withdraw. More than 80 male teachers remained.[53]

In the period after World War II, the number of degrees and certificates was measured through a survey of primary school teachers. Here it was shown as well that women had fewer additional degrees, and that they tended to receive further training in very specific areas, for example, in a second language or special education. By contrast, one out of every three men obtained, over the course of his career, a degree from one of the Higher Institutes of Educational Sciences.[54]

In Belgium, women teachers have not formed independent professional organizations as they have, for example, in the Netherlands;[55] instead, their interests are represented by unions organized around a philosophy. In general, the degree of union membership is lower for women teachers.[56] Whatever the case, the women who belong to unions are, in large measure, underrepresented in its leadership. Unions, however, have strongly supported its women members in their struggle for equal pay and helped to eliminate the rule prohibiting women teaching in Catholic schools from marrying.[57]

Regarding the number of school text books written by women and their participation in pedagogical journals, we once again find that there are no relevant studies available for Belgium. Fragmentary data exists showing that both in the nineteenth and twentieth centuries women did publish, but the number of women authors must have been much lower than that of their male counterparts.[58]

The extent to which women in teaching have generated innovations in the field has yet to be seen. But what is true is that they have not made their mark from crucial positions within the unions, or in-service training, or as authors of pedagogical works or textbooks. However, when compared to other professions, they have indeed played a leading role. Overall, the opening of professions to women has been most successful in the field of education, and more specifically, in the teaching profession.

THE RECENT MASCULINIZATION OF ADMINISTRATIVE POSITIONS AND THE FEMINIZATION OF PART-TIME WORK

The minority representation of women in positions of authority is, in general, a widespread phenomenon. Even when the nonprofit sector (which is markedly feminized) is included, in 1995 women held a scant one out of four high-level positions in businesses. In this respect, education is no exception.[59] Although the educational sector has had a strong female participation, it is disconcerting to observe that here, too, despite their high numbers, women are rarely promoted to administrative and/or executive positions.[60] Moreover, in recent years, we can speak of a masculinization of administrative positions.[61] It is striking to note how this masculinization has been accompanied by a change in responsibilities. In the past, principals were mainly involved in pedagogical matters, whereas today they are almost exclusively devoted to administrative duties.[62]

Until the 1970s—and even in the nineteenth century—there was a better equilibrium in the proportion of men and women: approximately

60 percent of primary school principals were female. In the following decades, the number of women primary school principals has decreased, which parallels a similar tendency in other levels of education.[63]

In general, the trend toward masculinization is most noticeable in the Catholic education system overall. Originally, the Catholic education system had more women principals because schools for girls were headed by women, who were frequently nuns. But the general criteria for the position of principal in the schools, which moreover had been reduced in number through rationalization, seemed too simplistic: a male principal in schools for boys and coed schools, and a female principal in schools for girls.

The unequal representation of women is not only seen in school administration. Differences can also be seen in the areas of state inspection and staff development services.[64] In primary, secondary, and training schools, the number of women working in 1989 in school inspection or staff development reached 21 percent, 6 percent, and 2 percent, respectively. The general trend is clear: in Belgium, education is considered to be a women's profession, but it becomes more masculinized as one ascends the educational hierarchy, or as duties or job positions become more policy-oriented.

In the same way, the feminization of part-time work constitutes a recent phenomenon. Part-time work in basic education has been available since the early 1980s, and women teachers above all have made use of this option.

Part-time work is not only dominated by women in the field of education.[65] As figure 6.9 shows, in 1988, one out of every five women worked part-time in Belgium, compared with one out of every fifty men.[66] Separate studies indicate that part-time work does not only mean that hours on the job are shorter; it also means, among other things, that one's job function is inferior and that opportunities for promotion do not exist. For these reasons, it gives rise to inequalities and reinforces segregation by sex.[67] In the field of education, access to a principal's position is more difficult if one holds a part-time job.

On the other hand, research studies show that the possibilities of part-time work, flex-time, and interrupted careers also motivate women to search for these jobs, for example, in high tech companies. And in the same way, in universities, it seems that the incompatibility between an academic career and a family lead women to abandon this field.

Is teaching perhaps the most favorable job when it comes to employer flexibility vis-à-vis the personal needs of the employee?[68] Women teachers are overrepresented among part-time workers.

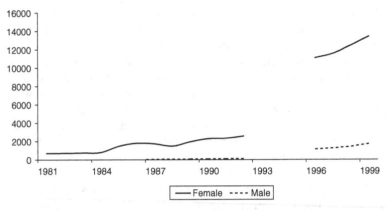

Figure 6.9 Total men teachers and women teachers who work part-time in Flanders (1981–1999)

Apparently, the educational structure appeals to women, and they very consciously choose it in order to reconcile the varied compromises and obligations in their lives. This may explain why teaching attracts women above all, despite the fact that proportionately fewer of them choose to study education.

In a recent study based on her *Preference Theory*, Catherine Hakim proposes that women *grosso modo* choose between three different styles of life and work: "*home-centered women*," "*work-centered women*," and "*adaptive women*" (for whom family and work are compatible).[69] These "styles" exist at all levels of teaching and in all social classes. *Preference Theory* predicts that the predominance of men in the labor market, in politics, will continue to exist, given that only a minority of women are fully centered on their job. But do women, and by extension people, live in a vacuum where they only make rational choices? The historical overview presented here, leads us to other conclusions.[70]

The feminization of the teaching profession and the masculinization of positions of authority, as expressed in quantitative terms, are processes that in our opinion, are closely related to the social phenomenon of modernization. With respect to this type of monolithic category, there is still, however, no thorough explanation. At the most, feminization (or as the case may be, masculinization) articulates, as has been considered here, one out of many dimensions of a deeper dynamic. It seems important, therefore, to affirm that the qualitative factors that we have passed over in our study may point the way toward a better explanation that, in the future, will also need to be corroborated with empirical data.

Our hypothesis is that the feminization of the teaching profession cannot be studied without taking into consideration the growing professionalization of the educational sector in society. And even more so when professional competence in this field seems to coincide more and more with characteristics that are considered to be typically feminine. After all, what is pedagogically correct was always synonymous with an emotional approach, gentle but firm.[71] Women are socialized with a sensibility that is compatible with the pedagogical vision, or to express it another way: the pedagogical constitutes the nucleus of the so-called femininity of women.

Perhaps we can make an historical analogy with the "G" (general) factor that Spearman[72] believed he identified in intelligence. As an attractive hypothesis for a more detailed study, we dare to affirm that the feminization of the teaching profession is closely related to the presence of a culturally and historically determined factor "F," which represents the social process of the "feminization of a woman." The general feminization of the female gender is "taught" to girls during their education because from various points of view it seems "pedagogically" correct. Feminization not only reflects and legitimizes a division of roles between men and women that is socially acceptable (the *hard* sector against the *soft* sector), but also discovers the essence of the so-called feminine nature in the caring power of "soft" professions (to which education unquestionably belongs). For this reason, as the influence of education grew—a process that can be defined as pedagogization, and which is, at the same time, a manifestation of the more general process of modernization[73]—the feminization of the teaching profession also grew, along with the feminization of education.[74]

In our opinion, this affirmation is given meaning by the fact that the process of feminization, like the process of modernization and of pedagogization, is a phenomenon that on the one hand has spread worldwide (a world that, as is well known, has been globalized as a function of modernization), and on the other hand, because of the clear advances in the emancipation of women, shows no signs of stopping.

Even today, such causes and effects of feminization, which refer to the heart of the matter of educational processes, can scarcely be traced in the professional literature on the history of female teachers.[75] An exception may be Juliane Jacobi's conceptualization of feminization as part of the modernization process.[76] However, that the refreshing exploration of the internal dynamics of everyday pedagogy can throw new light on such complex societal and sociological phenomena as the feminization of the teaching profession goes without saying.

NOTES

1. Finkelstein, B., "Foreword," in Rury, J. L., *Education and Women's Work. Female Schooling and the Division of Labor in Urban America, 1870–1930* (Albany: State of New York Press, 1991), p. XV.
2. Depaepe, M., De Vroede, M. and Simon, F. (eds.), *Geen trede meer om op te staan. De maatschappelijke positie van onderwijzers en onderwijzeressen tijdens de voorbije eeuw* (Kapellen: Uitgeverij Pelckmans, 1993).
3. The same lack of research is referred to by Gubin, E., "Libéralisme, féminisme et enseignement des filles en Belgique aux 19e-début 20e siècles," in Maerten, F., Nandrin, J. P., and Van Ypersele, L., *Politique, imaginaire et éducation. Mélanges en l'honneur du professeur Jacques Lory*, Centre des Recherches en Histoire du Droit et des Institutions, Cahiers 13–14, Facultés Universitaires Saint-Louis (Bruxelles: 2000), pp. 151–174. The most recent research study on gender in education in Belgium was published in 2000. Hermans, A., "Twee vleugels. Een verkenning van de thematisering van sekse in het historisch onderzoek over opvoeding en onderwijs" in Van Essen, M. et al. (eds.), *Genderconcepties en pedagogische praktijken (Jaarboek van de Belgisch Nederlandse Vereniging voor Opvoeding en Onderwijs)* (Assen: 2000), pp. 41–88. For an overview of women's studies in Belgium, in which education is also treated, see De Metsenaere, M., Huysseune, M. and Scheys, M., "Gewapend met het gewicht van het verleden: enige resultaten van vrouwengeschiedenis in België," in Duby, G. and Perrot, M. (eds.), *Geschiedenis van de vrouw. De twintigste eeuw* (Amsterdam: 1993), pp. 523–556.
4. Minten, Luc et al., *Les statistiques de l'enseignement en Belgique. L'enseignement primaire* appeared in four volumes between 1991 and 1996 and examines the period 1830–1992. This study was a product of quantitative analysis which we began at the end of the 1970s and out of which arose an interest in studying feminization as a quantitative phenomenon. See Depaepe, M., "Kwantitatieve analyse van de Belgische lagere school (1830–1911)," *Belgisch tijdschrift voor nieuwste geschiedenis*, 10, 2 (1979): 21–81; Depaepe, M. with the collaboration of Servaes, A., "Evolutie van het aandeel onderwijzeressen in het Belgische lager onderwijs sedert 1830," *Christene School. Pedagogische Periodiek*, 95 (1988): 204–210; Depaepe, M., "The Feminization of Primary School Teaching in Belgium" in Seppo, S. (ed.), *The Social Role and Evolution of the Teaching Profession in Historical Context. Vol. V: Male and Female Teachers in the History of Education, The Process of Feminization of the Teaching Profession* (Joensuu: University of Joensuu, 1988), pp. 96–105.

5. The interviews were conducted within the framework of a prior project. For an analysis of the interviews, see De Graeve, B. et al., "Het onderwijzersberoep en intergenerationaliteit. Een vier-generatie-onderzoek in Vlaanderen en Nederland op basis van oral history," *Tijdschrift voor sociale geschiedenis*, 11 (1985): 324–348. The interviews were also used in Depaepe, M. and Simon, F., "Social Characteristics of Belgian Primary Teachers in the Twentieth Century," *Cambridge Journal of Education*, 27, 3 (1997): 391–404.

6. Van Mechelen, R., *De meerderheid een minderheid. De vrouwenbeweging in Vlaanderen: feiten, herinneringen en bedenkingen omtrent de tweede golf* (Leuven: Van Halewyck, 1996) 120.

7. Among others, see Weiler, K., "Reflections on Writing a History of Women Teachers," *Harvard Educational Review*, 67, 4 (Winter 1997): 635–657; Weiler, K. and Middleton, S. (eds.), *Telling Women's Lives: Narrative Inquiries in the History of Women's Education* (Buckingham: Open University Press, 1999); Copelman, D.M., *London's Women Teachers. Gender, Class and Feminism 1870–1930* (London-New York: Routledge, 1996); For a recent survey of different Anglo-American studies on the historiography of gender and education, we refer to several articles in a special issue of *History of Education*, 29 (2000): 5, Theobald, M., "Women, Leadership and Gender Politics in the Interwar Years: the Case of Julia Flynn," 63–77, Goodman, J. and Martin, J., "Breaking Boundaries: Gender, Politics, and the Experience of Education," 383–388, Edwards, E., "Women Principals, 1900–1960: Gender and Power," 405–414.

8. Depaepe et al., *Geen trede*, pp. 43–44; Van Essen, M., "Strategies of Women Teachers 1860–1920: Feminization in Dutch Elementary and Secondary Schools from a Comparative Perspective," *History of Education*, 28, 4 (1999): 414.

9. Depaepe, M. and Simon, F., "Social Characteristics," 392; Albisetti, J., The Feminization of Teaching in the Nineteenth Century: a Comparative Perspective, *History of Education*, 22, 3 (1993): 253–263.

10. Benschop, Y., Brouns, M. and De Bruijn, J., "De kikker op het witte paard. Ontwikkelingen in het denken over sekse en gender in arbeid en organisaties," *Tijdschrift voor genderstudies*, 1, 2 (1998): 18; Vanderhoeven, J.L., "Masculinisering van beleidsfuncties in het Vlaamse onderwijs? Een verkenning," *Tijdschrift voor onderwijswetenschappen*, 21 (1990): 76.

11. De Weerdt, D., *En de vrouwen? Vrouw, vrouwenbewegingen en feminisme in België, 1830–1930* (Gent: Masereelfonds, 1980), p. 18; Peemans/Poulet, H., *Femmes en Belgique, XIXe-XXe siècle* (Bruxelles: Universiti des femmes, 1991), p. 27.

12. Hilden, P.P., *Women, work and politics: Belgium 1830–1914* (Oxford: Clarendon, 1993), pp. 3–7.

13. Keymolen, D., *Beroepsarbeid voor vrouwen van betere stand in België omstreeks 1860. Een bijdrage tot de mentaliteits- en onderwijsgeschiedenis*

(Gent: CSHP, 1981), p. 5; Minten, L. et al., *Les statistiques de l'enseignement en Belgique*. Vol. I: *L'enseignement primaire 1830–1842* (Bruxelles: Archives générales du Royaume, 1991), pp. 2–5.

14. Bracke, N., "De vrouwenarbeid in de industrie in België omstreeks 1900. Een 'vrouwelijke' analyse van de industrietelling van 1896 en de industrie- en handelstelling van 1910," *Belgisch tijdschrift voor nieuwste geschiedenis*, 26, 1–2 (1996): 166–167.

15. Vanhaute, E., "Modellen, structuren en de historische tijd. Op zoek naar een Belgisch kostwinnersmodel," *Brood en rozen. Tijdschrift voor de geschiedenis van sociale bewegingen*, 6, 1 (2001): 33–35.

16. Mitchell, B.R., *International Historical Statistics, Europe 1750–1993*, (London/New York: Macmillan, 1998), p. 136.

17. Pott-Buter, Hettie A., *Facts and Fairy Tales about Female Labor, Family and Fertility. A Seven-Country Comparison, 1850–1990* (Amsterdam: 1993), p. 21. Other statistics reflect a similar trend: Vanhaute, E., "Modellen, structuren en de historische tijd," 38; Matthijs, K., *De mateloze negentiende eeuw. Bevolking, huwelijk, gezin en sociale verandering* (Leuven: Universitaire Pers, 2001), p. 67; Lambrechts, E. *Vrouwenarbeid in België. Het tewerkstellingsbeleid inzake vrouwelijke arbeidskrachten: 1930–1972* (Brussel: Centrum voor Bevolkings en Gezinsstudiën, 1979).

18. Depaepe, M. and Steverlynck, C., "Op zoek naar de wortels van de vervrouwelijking in het onderwijs." Aantekeningen bij het tiende congres van de ISCHE te Joensuu (Finland) van 25 tot 28 juli, *Pedagogisch Tijdschrift*, 14, 3 (1989): 207–210; Depaepe, Marc, *De pedagogisering achterna. Aanzet tot een genealogie van de pedagogische mentaliteit in de voorbije 250 jaar* (Leuven: Acco, 1998), pp. 236–242.

19. Bracke, "De vrouwenarbeid," 174–200.

20. Wyns, M. and Van Meensel, R., *De beroepensegregatie in België (1970–1988)* (Leuven: HIVA, 1990), pp. 32,102, 108.

21. Schouwenaars, M., *Wat zal ons meisje worden?* (Antwerpen: t'Groeit, 1934).

22. Approximately two thousand teachers were conscripted. See De Clerck, K., De Graeve, B. and Simon, F., *Dag Meester, goedemorgen zuster, goedemiddag juffrouw. Facetten van het volksonderwijs in Vlaanderen (1830–1940)* (Tielt: Lannoo, 1984), p. 115; Bonet, V., *Loi sur la milice et sur la rémunération en matière de milice. Commentées et annotées* (Gand: Meyer-Van Loo, 1903), pp. 122–123 (art. 28 of the law of March 21, 1902).

23. Depaepe et al., *Geen trede*, p. 54.

24. Christens, R., *Onderzoek naar de ontwikkeling van de plaats en betekenis van meisjes en vrouwen in het onderwijs. 2. Onderwijzeres worden. Vrouwen tussen opleiding en beroep in de tussenoorlogse periode* (Leuven: KUL-Departement pedagogische wetenschappen. Afdeling historische pedagogiek, 1990).

25. Caperan, L., *L'invasion laïque: de l'avènement de Combes au vote de la séparation* (Paris: Desclee de Brouwer, 1935); Wynants, P., *Religieuses 1801-1975*. Vol. I: *Belgique–Luxembourg–Maastricht–Vaals*, Namur, Facultés Universitaires Notie-Dame de la, 1981 (Répertoires Meuse-Moselle, IV).
26. Tyssens, J., *Om de schone ziel van 't kind. Het onderwijsconflict als een breuklijn in de Belgische politiek* (Gent: Provincie Oost-Vlaanderen, 1998), p. 103.
27. Depaepe, M., De Vroede, M., Minten, L. and Simon, F., "L'enseignement maternel," in Grootaers, D. (ed.), *Histoire de l'enseignement en Belgique* (Bruxelles: CRISP, 1998), pp. 190-217.
28. Tihon, A., "Les religieuses en Belgique du 18e au 20e siècle. Approche statistique," *Revue belge d'histoire contemporaine*, 7, 1/2 (1976): 1-55.
29. Rogers, R., "Retrograde or Modern? Unveiling the Teaching Nun in Nineteenth-Century France," *Social History*, 23, 2 (1998): 147.
30. May, J., "Des 'religieuses dans le siècle' et des hommes de ce monde. Les élèves australiens de deux établissements d'enseignement secondaire non mixtes se souviennent de leurs professeurs (1930-1950)," *Histoire de l'éducation*, 98 (May 2003): 161-185.
31. The first female student to enter a Belgian university (Université Libre de Bruxelles) was admitted in 1880. In 1920 the Université Catholique de Louvain also opened its doors to female students. Despy-Meyer, A., "Les femmes dans le monde universitaire," in Mendes da Costa, Y. and Morelli, A. (eds.), *Femmes, libertés, laïcité* (Bruxelles: id. de l'Universiti de Bruxelles, 1989), pp. 47-58; Courtois, A., "Vers l'admission des étudiantes à Louvain (octobre 1920): jalons pour une histoire des mentalités catholiques en matière de condition féminine," *Revue d'histoire ecclésiastique*, 86, 3/4 (1991): 324-346.
32. *Statistisch Jaarboek van het Vlaams onderwijs*, 1998-1999.
33. Kabinet van de Secretaris van Maatschappelijke Emancipatie, *Vrouwen in de Belgische Samenleving* (Brussel: INBEL, 1991); Vanderstraeten, R., "De evolutie inzake onderwijsdeelname: naoorlogse tendensen en prognoses," *Tijdschrift voor Onderwijsrecht en Onderwijsbeleid*, 7, 1 (1996-1997): 21-29.
34. Leirman, W., "Het project onderwijs voor meisjes: van opinie- naar actie- en evaluatieonderzoek 1968-1973," *Tijdschrift voor Opvoedkunde*, 17, 5 (1971-1972): 276-304.
35. Depaepe et al., *Geen trede*, p. 43.
36. Lundgreen, P., "Die Feminisierung des Lehrerberufs: Segregierung der Geschlechter oder weibliche Präferenz. Kritische Auseinandersetzung mit einer These von Dagmar Hänsel," *Zeitschrift für Pädagogik*, 45, 1 (1999): 121-135.
37. *Statistisch Jaarboek van het Vlaams Onderwijs*, 1998-1999.
38. Vandenbroeke, C., "Het Lerarenambt als knelpuntberoep," *Vacature* (February 10, 2001): 3.

39. Depaepe et al., *Geen trede*, pp. 191–216.
40. Bracke, "De vrouwenarbeid," 199; Keymolen, D., *Vrouwenarbeid in België van ca. 1860 tot 1914* (Leuven: ACCO, 1977), p. 17; Creighton, C., "The Rise of the Male Breadwinner Family: a Reappraisal," *Comparative Studies in Society and History*, 38, 3 (1996): 310–337.
41. Depaepe et al., *Geen trede*, pp. 200, 207.
42. In Great Britain it was also teachers who took the initiative in the fight for an equal salary. See Pasture, P., "Feminine Intrusions in a Culture of Masculinity," in Pasture, P., Verberckmoes, J. and De Witte, H., *The Lost Perspective? Trade Unions between Ideology and Social Action in the New Europe.* Vol. II: *Significance of Ideology in European Trade Unionism* (Aldershot: Avebury, 1996), p. 223.
43. Depaepe et al., *Geen trede*, pp. 213, 221–235.
44. Hänsel, D., "Wer ist der Professionelle? Analyse der Professionalisierungsproblematik im Geschlechterzusammenhang," *Zeitschrift für Pädagogik*, 38, 6 (1992): 877.
45. Ozga, J. and Lawn, M., *Teachers, Professionalism and Class: a Study of Organized Teachers* (London: Falmer, 1981), p. VII; Lawn, M., *Modern times? Work, Professionalsm and Citizenship in Teaching* (London/Washington,: DC Falmer, 1996) p. 11.
46. Kemnitz, H., *Lehrerverein und Lehrerberuf im 19. Jahrhundert. Eine Studie zum Verberuflichungsprozess der Lehrertätigkeit am Beispiel der Berlinischen Schullehrergesellschaft (1813–1892)* (Weinheim: Deutscher Studien Verlag, 1999), p. 15.
47. Lawn, *Modern Times*, p. 62.
48. Ozga and Lawn, *Teachers, Professionalism and Class*, p. 35.
49. Nóvoa, A., "The Teaching Profession in Europe: Historical and Sociological Analysis," in Seing, E.S., Schriewer, J. and Orivel, F. (eds.), *Problems and Prospects in European Education* (Westport, CT: Praeger, 2000), p. 47.
50. Hänsel, "Wer is der Professionelle?," p. 878.
51. In her study Rebecca Rogers also shows that the ideal novice had to have a docile character: Rogers, "Retrograde or Modern?," 152.
52. Depaepe et al., *Geen trede*, pp. 120, 122.
53. From an interview with Julia Van Raetingen-Hendrickx (born in 1906). She taught between 1925 and 1958 in different Catholic schools. In 1957, she married at the age of 51.
54. Depaepe et al., *Geen trede*, p. 18.
55. Van Essen, *Strategies*, pp. 413–433.
56. Depaepe et al., *Geen trede*, p. 21.
57. Van Rompaey, L., *Strijd voor waardering. Het COV van 1893 tot 1983* (Antwerpen/Apeldoorn: Garant, 2003).
58. Sury, C., *Bibliographie féminine belge. Essai de catalogue des ouvrages publiés par les femmes belges de 1830 à 1897* (Bruxelles: 1898), cited in

De Weerdt, *En de vrouwen?*, pp. 28, 34; Depaepe et al., *Green trede*, pp. 25–26.

59. D'Hoker, N. et al., *Vrouwen in de Belgische Samenleving. Statistische gegevens 1970–1998*, Saint-Martens-Latemr: Van Eetvelde, 1998, p. 159.

60. For innovative studies on women principals in education, we refer to the special issue "Re-Examining Feminist Research in Educational Leadership," *International Journal of Qualitative Studies in Education*, 13 (2000): 6.

61. Vanderhoeven, *Masculinisering van beleidsfuncties*, p. 81.

62. Christens, R., Cuyvers, L., Gypens, V. and Steverlynck, C., *Onderzoek naar de ontwikkeling van de plaats en betekenis van meisjes en vrouwen in het onderwijs. I. Roldoorbrekende tendensen in de leermiddelen en in de onderwijsleersituatie* (Leuven: KUL-Departement pedagogische wetenschappen. Afdeling historiche pedagogiek, 1990), p. 19.

63. Depaepe et al., *Geen trede*, p. 53.

64. Vanderhoeven, *Masculinisering van beleidsfuncties*, p. 82.

65. McRae, S., *Le travail à temps partiel dans l'Union européenne: Les femmes et le travail à temps partiel* (Luxembourg: Office des publications officielles des Communautés européennes, 1996).

66. Wyns and Van Meensel, *De beroepensegregatie*, p. 44.

67. Ibid.; Hewitt, P., *About Time. The Revolution in Work and Family Life* (London: IPPR, 1993).

68. Lundgreen, *Die Feminisierung des Lehrerberufs*, p. 133.

69. Hakim, C., *Work-Lifestyle Choices in the 21st Century: Preference Theory* (Oxford: Oxford University Press, 2000).

70. See: Crompton, R. and Harris, F., "Explaining women's employment patterns: 'orientations to work' revisited," *British Journal of Sociology*, 49 (1998): 118–136.

71. Depaepe, M. et al., *Order in Progress. Everyday Educational Practice in Primary Schools. Belgium 1880–1970* (Leuven: University Press Leuven, 2000).

72. Spearman, C.E., *The Nature of "Intelligence" and the Principles of Cognition* (London: Macmillan, 1923), and *The Abilities of Man: Their Nature and Measurement* (London: Macmillan, 1927).

73. Depaepe, *De pedagogisering achterna*, pp. 13–31.

74. A similar trend in which the feminine element prevails in education is described in: "Delfos, M.F., De ontmannelijking van de man. Opvoeden met de biologische stroom méé in plaats van tegen de biologische stroom in," *Pedagogiek in Praktijk Magazine*, 7, 1 (2001): 8–13.

75. Van Essen, M. and Rogers, R. (eds.), "Les enseignantes XIXe-XXe siècles", *Histoire de l'éducation*, 98 (May 2003), special issue.

76. Jacobi, J., "Modernisierung durch Feminisierung? Zur Geschichte des Lehrerinnenberufes," *Zeitschrift für Pädagogik*, 43 (1997): 929–946.

PART III

The Feminization of a Profession at the Turn of the Twentieth Century

CHAPTER 7

Women and Teaching in Costa Rica in the Early Twentieth Century

Iván Molina

In 1913 the writer Rómulo Tovar published an essay in defense of Mauro Fernández, the Minister of Public Instruction who had been the driving force behind the educational reforms of 1886 that were aimed at centralizing and secularizing primary education in Costa Rica.[1] The text, however, warned with certain misgivings:

> [T]he problem with education is at present expressed in almost the same terms as it was twenty-five years ago: the teacher in general is deficient. We do not wish to ignore the good qualities of some highly recommended educators, but these form no more than a select minority. The remainders are unprepared or hastily prepared, with no ambition, unaware of the work they are performing and without initiative, and as they are in the majority, in the long run they will exert the largest influence over the public spirit.[2]

Tovar's opinion was supported by the evidence. In 1911—the year closest to 1913 for which information is available—only 13.5 percent of all teachers (129 out of 953) had a diploma.[3] The low diploma rate, however, was not the only matter troubling education authorities. Indeed, its main concern at the beginning of the twentieth century was the gender of teachers, and in particular the increasing feminization of the teaching field. The purpose of this essay is to explore this issue in terms of its market dynamics (noting characteristics specific to urban and rural areas, family circumstances, chances for promotion, among other subjects) using a little-known source.

The Costa Rican government published on June 25, 1904 a register entitled *The Organization of Teaching Personnel in Primary Schools*, which is an authentic—although limited—census of men and women who gave classes in schools throughout the entire country in the specified year.[4] The main advantage of this official census is that it identifies teachers by name, indicates those who occupy the position of principal, locates schools by province, canton, and district, and in the case of women teachers, includes a designation by which their marital status can be inferred (married, single, or widowed).

The inclusion of names along with the information offered by this register is useful in identifying the teachers who were in charge of special subjects (such as singing, drawing, religion, sewing, and gym), those who were married, and the men and women who most likely had some sort of family relationship (e.g., two female teachers in the same school with an identical last name, or a couple who taught classes together with an unmarried woman with the same last name). The limitations of such data are obvious and make for a certain lack of precision; but nevertheless, a rough idea can be formed of some of the basic patterns that characterized the teaching market at the beginning of the twentieth century.

The census of 1904 was done at the beginning of an era when Costa Rica underwent important political and social changes: once the authoritarian regime of Rafael Iglesias (1894–1902) had ended, there was a decisive democratic opening during which the growing competition between political parties allowed the urban and rural popular classes to channel their demands into the electoral process. The principal effects of this dynamic were soon visible in the creation of new cantons or municipalities (twenty-two between 1903 and 1915),[5] which was in response to the pressure of local power hierarchies, and in the reorientation of state spending. The proportion of the national budget that was designated for education, health, pensions, and public works (which also included infrastructure for schools and sanitation), increased from 24.4 percent to 34.3 percent between 1890–1901 and 1902–1916; and in those same years, the annual growth rate of public employment increased from 2.1 percent to 4.1 percent.[6]

The dictatorship of the brothers Federico and Joaquín Tinoco (1917–1919), which replaced the social-reformist government of Alfredo González Flores (1914–1917), interrupted these trends for a short time; but they deepened during the decade of the 1920s when the funds destined for education, health, and public works reached 38.8 percent of state expenditures. The national census of 1927 allows us to appreciate, in turn, how the impact produced by the increased governmental spending was differentiated by gender. Although more men were employed (5,969) than women (2,762), the former represented scarcely 4.4 percent of the men in the population who were economically active, while the latter represented 16.5 percent of economically active females (more than half of that proportion, 8.6 percent, were teachers and professors).[7]

The degree to which political democratization favored an early increase in public employment among females is a subject that has still been little investigated. A specific research study is unquestionably needed; but at present, what does indeed stand out is that the growing insertion of women into the government labor market was opposed, without any delay, by men. In particular, intellectuals and professionals, for whom the State was a strategic source of jobs and income, didn't hesitate to denounce the invasion of a workspace they had defined as theirs, a process whose most visible manifestation was the feminization of the teaching field.

SCHOOLS AND TEACHERS

Table 7.1 shows a geographic distribution of teachers and schools that goes beyond a strictly provincial perspective by classifying the former and the latter categories according to location in cities (capitals of provinces), towns (seats of cantons) and rural areas (the rest of the

Table 7.1 Geographic distribution of schools and teachers in Costa Rica and average number of teachers per school (1904)

| Province | Cities | | | Towns | | | Rural Area | | |
	Schools	Teachers	Average teachers	Schools	Teachers	Average teachers	Schools	teachers	Average teachers
San José	8	108	13.5	30	105	3.5	42	73	1.7
Alajuela	3	38	12.7	12	58	4.8	78	127	1.6
Cartago	3	33	11.0	8	41	5.1	31	55	1.7
Heredia	3	36	12.0	16	67	4.2	21	37	1.8
Guanacaste	2	19	9.5	9	28	3.1	22	32	1.5
Puntarenas	2	19	9.5	6	20	3.3	12	16	1.3
Limón	2	4	2.0						
Total*	23	257	11.2	81	319	3.9	206	340	1.7

* Includes a mixed school in the city of Cartago, one in the town of San Ramón in the province of Alajuela and 65 in rural areas.

Source: Costa Rica, *Organización del personal docente de las escuelas primarias* (San José: Tipografía Nacional, 1904).

canton). The second of these categories includes, however, the schools located in various districts that, in the near future, would achieve cantonship or would acquire an urban aspect that was slightly more defined.[8] This exception is explained by the fact that since 1904, the schools of those jurisdictions had characteristics that distinguished them from schools in the countryside.

The main contrast that can be seen in Table 7.1 is between the urban and rural school infrastructure: in effect, 7.4 percent of schools are concentrated in cities and 28.1 percent of teachers, for an average of 11 teachers per school. The figures were lower for the capitals of Guanacaste, Puntarenas and above all of Limón (a province that had only two public educational centers in 1904); but, in its entirety, the urban teaching establishments were noted for their better structure and organization than those in towns and in the countryside.

Rural Schools

The difference between schools of the first rank (which offered all grade levels, from first to sixth), schools of the second rank (which ended at the fourth grade), and schools of the third rank (with the first two elementary grades), is decisively linked to the level of urbanization in the communities they served. The reasons behind this hierarchization of the schools and of the teaching provided in their classrooms was explained in detail by F. F. Noriega, an inspector in the province of Alajuela, in April, 1904:

> [W]e should begin with the assumption that in the countryside we are going to teach mostly men who perform agricultural work, among whom 95 percent urgently need to receive the largest amount of knowledge in the shortest time possible. This knowledge should be restricted to the most practical, and we should forego the maximum culture we pursue for the inhabitants of the towns and cities, where a certain intellectual gymnastics is more necessary; it must be replaced by a larger amount of practical knowledge. We should consider that education in the countryside is for rough, practical laborers, and in the cities and towns, for those aspiring to public positions and liberal professions.[9]

If we examine the information in the 1904 register, Noriega's focus can be widened in various ways: among the 206 rural schools opened that year, in 75 the instruction depended on one individual; in 128 two people worked; in only three schools did three teachers give classes. However, the number of education by one individual schools was greater than these numbers reflect, since in 112 of those 128 educational centers, a man teacher instructed the boys and a woman teacher instructed the girls. So the proportion of these kinds of schools increased to 90.8 percent in the countryside.

Rural schools comprised, in turn, the majority of the mixed schools (for boys and girls): of the 67 schools of this type that existed in 1904, one located in the urban center of Cartago was staffed by 11 people, another, staffed by seven educators, was in the town of San Ramón, and 65 were spread throughout the countryside. In the latter schools, 49 were staffed by only one person and in only 16 was the teaching shared by two individuals. What this means in practice was that in 75.4 percent of the schools in the rural universe (more than three-quarters of them), the combined instruction for boys and girls depended on a single teacher.

The mixed schools in Cartago and San Ramón were, to a certain degree, the exception, as they were first rank schools, located in urban spaces, in which all grades were offered. Third rank schools, in which only the first two grades were offered, were a response to the dispersion of the rural population, which, in the case of Costa Rica, was a constant factor, fueled by the agricultural colonization carried out by peasants during the nineteenth and twentieth centuries. Their search for virgin land, especially between 1830 and 1950, continually displaced families from existing school infrastructures.[10]

Urban Schools

School infrastructure in cities, with the exception of Limón, was very different: of a total of 21 schools, there were between seven to nine teachers at 11 schools, between 10 and 12 teachers at seven schools, and in three schools, between 13 and 14 teachers. These differences, given that all were schools with grades one through six, can be explained once again by the size of the population served and by the fact that teaching was concentrated in a few, large buildings (the most notable example of this process was the Metálico Building, in San José). The urban schools, moreover, accounted for 49 of the 57 teachers of special subjects who worked in Costa Rican primary school education.

The range of qualitative variation between the city schools (all first rank) and those in rural areas (of which the majority were third rank), was limited when compared to towns. The differentiated urban development of those small towns—which served as administrative, commercial, and political centers within rural surroundings—left its mark in the 1904 text: in ten of 83 school establishments, one or two individuals were in charge of teaching; in 47, there were between three and four teachers, and in 24, between five and eight teachers.

According to the data, 12.3 percent of the schools in towns resembled the rural schools, while 29.6 percent of these schools had conditions similar to those in cities. This wider variation among schools in towns reflected, without a doubt, with the size of the population and with its degree of urbanization; but also, and most importantly, the wealth and political influence of its residents (and the differentiated importance of the local leaders in the provincial and national elections). This last factor explains why, in the canton capitals which were similar in terms of demography and urbanity, the educational infrastructure was so different.

THE FEMINIZATION OF THE TEACHING PROFESSION

The disparities between schools in cities, towns, and the countryside, together with the specificities of teaching in those spaces, affected the dynamics of the teaching labor market on several levels. The fundamental transformation experienced by teaching at the end of the nineteenth century, was the growing feminization of this profession, a subject that has not yet been adequately studied.[11] The 1883 census shows that in that year, on the eve of the 1886 reform, of the 241 teachers in the country, 105 (or 43.6 percent) were women. Men outnumbered women in all the provinces, and in the cantons in which men were not the majority, the proportions between men and women were roughly equal.[12]

The census data from 1892 confirm these tendencies: of the 451 teachers counted, 199 (44.5 percent) were women, and once again, male teachers predominated in the provinces, and in only 10 of 31 cantons was there an equal number of men and women teachers. The male advantage, however, disappeared over the course of the following four years: in 1896, of 784 teachers, 447 were women (57 percent).[13] The feminization of the profession was, in light of these statistics, a sudden phenomenon, which took place in a period of less than five years.

The participation of women in the labor market for teachers increased by 6 percent between 1864 and 1883, by 5.4 percent between 1883 and 1892, by 11.7 percent between 1892 and 1900, and by 16.5 percent between 1900 and 1915. Figure 7.1, despite the fragmentary nature of its data, demonstrates that the feminization of

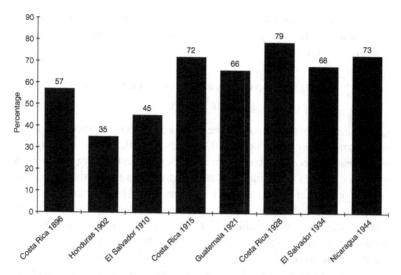

Figure 7.1 The feminization of the teaching field in Costa Rica and the rest of Central America: percentages of women teachers in primary education (1896–1944)

Source: The same as n. 14.

teaching occurred earlier in Costa Rica than in Honduras and El Salvador. The percent achieved by Costa Rican women teachers in 1915 exceeded, in turn, the rate of Guatemalan women teachers in 1921 and Salvadoran women teachers in 1934, and was almost the same as the percent achieved by Nicaraguan women teachers in 1944.[14]

The advances achieved by women caused considerable concern among Costa Rican educational authorities, and their discomfort was expressed in unmistakably military metaphors. The Inspector General of Education, Miguel Obregón Lizano, in April 1904, complained that:

> [M]en teachers represent 39.45 percent of the total, at the same time that women are at 60.54 percent. It is clear, then, that in this year [1903], as in previous ones, the female element is predominant among our teaching personnel. The causes of this growing invasion by women into teaching are well known to this Secretary [of Public Education].[15]

In 1904, Professor Obregón Lizano considered it unnecessary to explain what these "causes" were, but in 1915, when the rate of men teachers was at 27.7 percent, Luis Felipe González Flores, the

Minister of Education, was more explicit when he attributed to economic factors "the growing desertion of the masculine element in National Teaching."[16]

THE LABOR MARKET FOR TEACHERS
IN PERSPECTIVE

González Flores's explanation seems to be correct: the monthly salary of a man teacher went up to 50 colones in 1902;[17] and although at times teachers could increase their salaries by performing other work in school besides teaching, such as monitoring exams at the end of the school year, their nominal salary was not much higher than that of a skilled workman. Masons and carpenters earned between 50 to 70 pesos a month at the end of the 1880s, and at the same time it wasn't unusual for a more specialized worker, printers for example, to reach salaries close to or slightly above 100 pesos a month.[18]

Age complicated the salary question even more. An apprentice—a typical job category among laborers under twenty years old[19]—earned between 12 and 24 pesos a month; by contrast, the salary of a young man teacher in a grade school was double or triple that amount. The laborer, however, as he grew older and his skills increased, could double or triple his initial salaries, an expectation that a man teacher did not have, as promotions within the school organization were very limited. Primary school education, therefore, was an attractive entry-level option into the non-manual labor market for a young man with a certain degree of education; but it did not offer sufficient incentives to dedicate himself to education over the long run.

The case of women was distinct: although their salaries were less than those of men teachers (approximately 20 percent less between 1892 and 1902),[20] the gap was not as wide as the wage differential between women laborers and their male counterparts. The available information demonstrates that the wages of male laborers was two to three times as high as those of female laborers.[21] Teaching was an option that offered young women—apart from its being intellectual, and not manual—a job in which salary inequities were not as sharp as those in other occupations.[22]

The fact that the greatest salary differentials between men and women existed in the field of manual labor, and the least in the field of teaching, was a basic impetus for the vertiginous feminization of teaching, a more prestigious alternative on the labor market for a young woman who had no access to careers in law, pharmacy, or medicine.[23] A young man's perspective was different: working in education only made sense for a short time, given the lack of competitive salaries over the long run and the existence of diverse and attractive employment options. The degree to which the economic component was less important to women than men, a question posed in the now classic studies by Etzione and Lorti,[24] is a theme that remains open to future research; in any case, the available evidence for the first decades of the twentieth century makes it clear that women teachers fought tenaciously against salary discrimination, as will be seen later.

The State, lacking sufficient funds to deal in an appropriate way with the expansion of educational services that were demanded by the population, was the principal promoter of the feminization of teaching, since it opted to use cheap women's labor instead of accentuating the salary discrimination that existed by favoring men. The effect of such a policy was that the increase in educational employment, which took place in the decade of the 1890s, elicited an immediate response from women: of 170 new positions created in 1903, 112 (65.9 percent) were filled by women.[25]

The feminization of teaching occurred almost in a parallel fashion with the opening in 1888 of a teacher training school in the Liceo de Costa Rica, an institution for men, from which few men teachers graduated, and of another similar teacher training section in the Colegio Superior de Señoritas, from which an average of 12 women teachers graduated annually between 1891 and 1903.[26] A concern about finding sufficient applicants to be trained as teachers dates from 1849, when the first women's teacher training school was founded; however, this was a short-lived institution that closed in 1856. The State's commitment to develop educational personnel was only resumed 30 years later, within the context of the educational reform of 1886.[27]

The teacher's training section of the Colegio Superior de Señoritas gave degrees to an average of 19 women teachers per year between 1904 and 1914; in this last year, the section closed its doors, as a new public teacher's training school for men and women, under the coeducational system, was opened in the city of Heredia.[28] The women graduates from the Colegio, whose pedagogical experience is better known, received an education that was predominantly scientific and, in ideological terms, had a modern, feminist orientation. In the first decades of the twentieth century, these young women became the principal promoters and defenders of the social policies of the liberal State, especially in the field of health, at the same time they protested against salary discrimination and mobilized in favor of their right to vote.[29]

MALE DESERTION BY GEOGRAPHIC REGION

The desertion by men from primary school education was linked to the growth and diversification that urban economies experienced at the end of the nineteenth century.[30] Labor options for men, in that context, expanded and at the same time became concentrated in three basic areas: state institutions, whose employees doubled from 2,118 to 4,441 between 1881 and 1905;[31] commerce, energized by variations in patterns of consumption; and an early mass culture, which increased available employment in activities associated with leisure time—theater, films, sports, and other activities—and in print culture, especially periodicals (approximately 187 magazines and newspapers were published in the country from 1903 to 1914).[32]

Male desertion was not a particularly uniform process. It was more accentuated in San José, the axis of the country's urban economy with 24,534 inhabitants, than in other cities, and in the city of Liberia (in province of Guanacaste), men still outnumbered women in 1904. The abandonment by men was lower in Alajuela and Cartago than in Heredia, a difference that can be explained by its demographic weight (a population of 7,499 in Heredia compared to 5,502 and 5,626 in Alajuela and Cartago), and its proximity to San José. Puntarenas, in turn, because of its status as a port (although it was no longer the principal one in the country, and its population of 4,528 was less than other provincial capitals), offered attractive options for employment in the services sector.[33] Towns stood in visible contrast to cities, given the greater proportion of men on the teaching staff (in total, almost double the percentage in cities), particularly in San José, Heredia, and Guanacaste. The experiences of these towns prove that male desertion from primary education depended decisively on the options that became available within the urban economy: since job possibilities were more limited in the canton seats, the teaching profession maintained an advantage there that tended to disappear in provincial capitals, especially in the city of San José.

Rural areas, with the partial exception of the regions of Cartago and Guanacaste, experienced a high male desertion, which can be attributed to men teachers' interest in relocating, for example, from the countryside to towns or cities. Town centers or cities offered them a higher quality of life, in terms of services (including educational services, which were important for married teachers with children), and better consumer options and employment. The importance of the latter was related to the expectation of securing another job that was more lucrative than teaching, but also of ascending within the ranks of the educational system.

Promotion within the framework of the school system was greatly dependent on moving from place to place and from low-income positions to better paid ones. The salary for a teaching position in a rural mixed school with only one instructor was no higher than 40 colones a month; in contrast, a teacher in the boy's high school in the city of Alajuela earned 65 colones a month (62.5 percent more than in the rural school). The employment options that, in particular, attracted teachers were those of holding the principal's position (in the grade school in the town of Grecia the principal's monthly salary was 100 colones),[34] or in building a career in school administration (e.g., as an inspector).[35]

The main route to success for a male teacher consisted in holding a position in one of the prestigious schools in the city of San José, which enjoyed the best resources and infrastructure, and if possible, moving from primary to secondary school teaching, a transfer that opened up a whole spectrum of opportunities. The experience of the teacher Juan Dávila is a case in point: after graduating in 1901 from the Pedagogical Institute of Chile, in 1903 he earned 150 colones a month as a history teacher in the Colegio Superior de Señoritas (High School for Young Women); he added 130 colones to his monthly salary by teaching geography and cosmography in the Liceo de Costa Rica; he also taught commercial law and political economy in the School of Commerce, all of which gave him a monthly salary of more than 300 colones.[36]

The higher salaries paid in secondary school education explains why, with respect to the labor market, and in contrast to primary schools, they remained under a decisive masculine control. School statistics from 1904 demonstrate that, of the 29 teachers who worked at the high school level in the Liceo and in the Colegio Superior de Señoritas, only seven (24.1 percent) were women.[37] Their proportion in all the high schools in the country increased to only 34.8 percent in 1941 (54 out of 155 teachers); and in that same year, of the 94 professors on the faculty of the University of Costa Rica when it opened its doors, only six were female.[38]

The other job area in education that remained strictly controlled by men was serving as proctor for end-of-term exams at the primary and secondary school levels: of the 30 people chosen to carry out this job in 1903, whom it is possible to identify by name, only one was a woman. The participation of lawyers and doctors in this activity doubtlessly contributed to increasing, in total, the honorariums given to the delegates, among them teachers, since the rates were set in order to be attractive to those in the legal or medical field who, of course, had a higher income than a teacher. The principal of the school in the town of Aserrí, Maximino Blanco, earned 100 colones for proctoring the exams held in San José's schools that year, and at the same time, Manuel Benavides, a drawing teacher, earned 167 colones by performing a similar job in the city of Heredia.[39]

The female invasion into the educational system, which Obregón Lizano complained about in 1904, and the desertion by males, lamented by González Flores in 1915, were two processes that were more complicated than these educational authorities ever imagined, given their specific geographic and labor characteristics. The tendency was for men to abandon the areas of the job market for which the salary was very low (in the countryside) or when more attractive options for employment existed (in the cities); but they outnumbered women in the better paid jobs within the central educational hierarchy, among exam proctors, and in secondary school education.

Family Strategies and Teaching

The growth and diversification of the urban economy, which widened the occupational choices for men, and the increase in the demand for a low-wage teaching force, generated by the reform of 1886, favored the "invasion" of women into the educational system. This process, apart from the existing structural factors, was facilitated by a series of family practices, which helped consolidate the feminine insertion into primary schools. One of these strategies was the formation of teaching couples, which tended to arise because of the man teacher's interest in securing a job for his wife that would complement their domestic income.[40]

Unions of this kind were infrequent in cities, but had a greater importance in towns and in the countryside. The classic example was of a rural teacher working in a mixed school or a one-room schoolhouse and who, at a given moment (usually right before another position was opened), "offered" the educational services of his wife. This procedure may have displeased the school authorities, but they tolerated it, for the sake of avoiding the desertion by another man; moreover, the wife of a male teacher, given her condition as a potential mother-teacher,[41] might be a better candidate than another aspiring teacher who also had no experience and was without a degree.

FAMILY TIES AND THE LABOR MARKET

The initiative to place a wife in the school system did not always originate with the husband, and at times it was likely that the wife herself made the proposal. The work of women in teaching was, on occasion, offered free of charge, with the hope that later, after the community had begun to rely on her services, and if these had been deemed satisfactory (especially by the inspector), a formal salaried position would be created. From this point of view—employing the military vocabulary of González Flores and Obregón Lizano—male teachers not only deserted, but they were complicit with the enemy: in their eagerness to increase the household income, they became key allies to the feminization of the teaching profession.

The labor strategies from a domestic base were not limited to wives: in seven of the 36 couples who figured in the data from 1904, the husband and father, presumably, succeeded in placing not only his wife, but one of his sons or daughters in a teaching position. The placement of a wife, daughter or other female family member in the school system had, besides a family component, electoral implications as well: although women's suffrage in Costa Rica was only approved in 1949, it was difficult for political parties and politicians to ignore the demands of male voters anxious to add their wives to the state payroll, especially because many of these men were part of local power hierarchies. In this way, at the same time democracy contributed to the feminization of the teaching profession, it also limited the means that school authorities could use to halt this process, given its potential cost in votes.

Teaching couples were not the only ones who followed the above practices. The data about kinship is somewhat limited by the fact that family ties were inferred among teachers having the same last name who worked in the same school; as a result, it is possible that kinship ties were over counted when no family relations existed between the people to whom they were assigned. A corrective to this error, however, is also contained within the source data, as it undercounts family ties between cousins, nephews, nieces or in-laws, among others, whose last names are not always the same.

Despite these limitations, the tendency is clear. once a member of a family was integrated into the teaching profession, he sought to incorporate other members as well. In geographic terms, the dynamic of this process was distinct from that of a married couple: the proportion of the latter increased as the workplace became more rural; in the case of non-marital kinship, its frequency increased as the work environment became more urbanized. A brief methodological detour is needed to explain these contrasting forces at work in the domestic realm of the educational universe.

KINSHIP AND GENDER IDENTITY

Although the 1904 register allows one to infer kinship on the basis of last names, it is not helpful in determining the exact type of relationship. For example, in the case of two single women with the same last name who taught at the same school, they were perhaps sisters, but they could also have been cousins. The possibilities for men were still wider, since in addition to cousins, they could also have been father, son, uncle, and nephew. These difficulties made it necessary to limit the analysis to three basic criteria: if kinship ties were only between men, between women, or between men and women.

The specific relationship between family members, in light of these categories, can be broken down as follows: of the 224 cases in which the relationship was not marital, eight were children of married teachers; in 120, the family relationship was only between women, an indicator of the ongoing process of a female identity in the making; in 20, the relationship was only between men; and in 76, both men and women were involved. Women teachers, according to these figures, completely dominated 53.6 percent of the non-marital family ties, and they moreover were present in 33.9 percent of the relations which included a male teacher. This distribution, however, had its own characteristics across the rural–urban spectrum.

Family kinship, with a total predominance of women, reached 66 percent in cities, 48.5 percent in towns and 36.5 percent in the countryside. The tendency for the rate to decrease is in accordance with the earlier data: in small town centers and in the rural universe, labor markets where the importance of male teachers was greater, family ties to a male teacher were still strategic tools for a married or single woman to obtain a job in a school. The dynamic in provincial capitals was quite different, as women teachers were able to convert family ties that were specifically between women into a means to obtain a job, a process facilitated by male desertion.

The construction of a family tradition in teaching was, therefore, a strategy practiced especially by women: in contrast with the 53.6 percent of non-marital kinship ties among only women teachers, those existing between only male teachers accounted for only 8.9 percent of the total. The considerable advantage held by women in this area makes it evident that men, despite their higher numbers in primary school education during the nineteenth century, did not form a strong gender identity around working in schools, and this lack facilitated their subsequent desertion and the female invasion into the school system.

The level of family ties between men and women, which made up 33.9 percent of all instances of kinship, affirms the decisive role played by women teachers in constructing occupational traditions in education. The experience of the young man who obtained employment in a school thanks to the support of a woman relative already at work in education, was not an exception, especially when the relationship was between mother and son. More typical, however, was the reverse: married and single women who relied on a male teacher in their family to join the ranks of educators. Kinship thus contributed to the feminization of the teaching field and implied that this process, in an intertwined way, would result in strategies and family dynamics oriented toward searching for options in public employment for one or several members of the family, most often women.

The decrease in the rate of kinship ties as the workplace became more rural had less to do with the level of urbanization and more with the unequal influence of women educators in different labor markets. Women, as the decisive shapers of a family tradition in teaching, were especially successful in cities, where their numbers were higher as a result of male desertion.[42]

Taken as a whole, the frequency of kinship in the primary schools can be better visualized when to the 224 cases of non-marital kinship ties are added the 72 cases of married men and women. The total, 296 individuals, accounts for 32.3 percent of the teaching staff in Costa Rican schools in 1904. This fact, which underscores once again the crucial role played by family ties between women, is proof of the importance of kinship for the labor market and for the professional identity that existed in the teaching corps, which proved to be key in mobilizations like the mass protests in June, 1919 led by women teachers and professors against the dictatorship of the Tinoco brothers.[43]

POSITIONS AS PRINCIPAL AND THE MARRIAGE MARKET

The educational hierarchy was not unaffected by the male desertion and female invasion in the teaching field: a significant space was opened in this context which allowed women to fill positions giving them official power. The 1904 register reveals that, in this respect, there still existed a small male advantage: while they filled only 37.6 percent of the teaching positions, they held 42.8 percent of the principalships.

The majority of the latter positions were filled by women, although the rate did not match their participation in the labor market as teachers.

The Unequal Distribution of Principalships

Gender inequalities within the position of principal can be seen in the proportionately greater participation of men in different laboral spaces: in cities, they account for 27.6 percent of teachers and 30.4 percent of principals; in towns, these figures increase to 48.7 and 54.1 percent, respectively, and in rural areas to 34.7 and 35.3 percent, respectively. The small male advantage in the school hierarchy indicates that the expectation of holding a position of power was insufficient to avoid male desertion, which in turn provided greater opportunities for the promotion of women.

Access to principalships was conditioned by the particular characteristics in the different workplaces: in cities, all the schools had a specific position for a principal, except for the schools in Limón, where the person in charge of the school also had to teach. This was the predominant model in towns: 74 of their 81 schools (91.3 percent) were directed by a teacher with additional responsibilities;[44] only three had independent positions for principals, although their duties included administering two different schools, one for girls and one for boys; and in one school located in the seat of the canton of Bagaces, in Guanacaste, there was no principal's position at all.[45]

Rural areas were strikingly different: of 206 schools, only 68 (33 percent) had a position of principal-teacher. The significance of such a geographic imbalance in principal's positions was obvious: for teachers interested in ascending the hierarchy of the school system, and especially for male teachers, a special incentive existed to relocate from rural areas to towns. Although towns had only 26.1 percent of schools, 45.2 percent of all principals' posts were concentrated there. This discrepancy goes far toward explaining why towns were less affected by male desertion.

The average ratio of teaching positions per principal was, altogether, 11.2 in cities, 4.2 in towns and five in rural areas;[46] and specifically, for males alone, the respective ratios were to 10.1, 3.9, and 4.9 male teachers per principal. The growing male abandonment of primary school teaching was conditioned, therefore, on the unequal geographic distribution of the school hierarchy, which had an impact on the opportunities for increased salary and promotion.

The two least attractive areas in this respect were, in order of importance, the main urban centers of the country, and rural areas.

The desire to hold the post of principal arose not only because of the fact that its occupant would earn a higher salary than a teacher (at times, 50 percent more), but also because the gap between the salary of a male principal and a female principal was greater than the gap between male and female teachers. The financial advantages that favored men, in this case, could reach up to 20 percent; by way of contrast, in 1903, the principal of the boys' school in the town of Grecia had a salary of 100 colones per month, while his female counterpart at the girls' school in the same town had a monthly salary of 50 colones, a salary differential of a hundred percent.[47] The disparity between the salaries of male and female principals can be explained because the expectation of a promotion with a significant raise in salary was an incentive for men to remain in the school system.

The procedure and the criteria by which educational authorities promoted a particular person (man or woman) to the post of principal is a subject that awaits further investigation, as does the means by which both male and female teachers transferred from schools paying low salaries to those offering better remuneration. The degree to which family connections or political contacts could influence the mobility of male and female teachers is not known, but most likely it was significant, given the fact that in the first decades of the twentieth century, Costa Rica was a society on a small scale—in 1927 the population was 471,524[48]—in which personal encounters (face to face) were the rule.

MARITAL STATUS AND THE LABOR MARKET

The labor dynamic in primary education, besides the distribution of principals' positions, was also quite influenced by the matrimonial opportunities for men and women. The municipal census in the city of San José in 1904 provides several details that help address this question: of the 122 women educators counted, 102 were single, six were married, and 14 were widowed, with the two last categories accounting for 16.4 percent of the total. The experience of men teachers was very distinct: of 46 men, 27 were single, 16 were married and three were widowed, with these last two categories representing 41.3 percent.[49]

The 1904 register only allows one to determine the woman's marital status and not the man's. According to its data, of the 572 women teachers counted, 73.6 percent were single, 21.7 percent were married and 3.3 percent were widowed; it is impossible to ascertain the status of the remaining 1.4 percent. The variations in these percentages across diverse work settings, demonstrates that the tendencies that are evident in San José were similar to those found in the urban universe on the whole.

The data on marital status once again show the striking difference between the male and female experience. The overwhelming majority of single women stands in contrast to the male profile, as almost one-half of the men are either married or widowed. The percentage of educators with a similar marital status in towns and the countryside was probably higher than in San José, but there is still little information available. In any case, the province of Puntarenas offers useful information in this respect: of its 51 men teachers in 1903, 22 had a wife and two were widowed, which together accounted for 47.1 percent of the total, a rate higher than in the city of San José.[50]

The predominance of single women in primary education increased as their place of work became more urbanized, a tendency that might be related to the fact that in cities the competition for available positions was more intense, and young, single, educated women had an advantage over married women and widows. The low percentage of women in urban areas in the latter two categories can also be explained by the fact that women teachers who married tended to leave the field, a practice that was not as common in towns or in rural areas, where women teachers by tradition did not abandon the school system despite getting married or becoming widowed.

The information provided by the 1904 municipal census in San José corroborates what has already been shown: of the 102 single women teachers who were living in San José that year, 59.8 percent were between 15 and 24 years of age, 20.6 percent were between 25 and 29, and only 18.6 percent were over 30. This distribution of ages suggests that, by age 24, when the majority of the women teachers had already married, they tended to leave the field to devote themselves to household work; therefore, the incorporation of these women teachers into the teaching workforce was more of a short-term strategy than a long-term commitment.

Marital status also affected school dynamics at two basic levels, which were especially visible in cities. The desertion of young women who married delayed the professionalization of the teaching corps, as marriage removed from service a number of qualified teachers. This process, in turn, led to a constant high demand for women teachers, since it was necessary to replace those who left. The abandonment of teaching by older women and their substitution by younger teachers resulted in a loss of accumulated experience; but seen from another perspective, it facilitated the renewal of the teaching staff, as it promoted the systematic incorporation of young women who were up to date with respect to official values and knowledge.

The removal of married women from the teaching workforce was strongly linked to the limits that marriage imposed on their geographic mobility, a basic requirement in order to remain and be promoted in the school system. A man teacher, although married, could easily relocate from one place to another, because of his status as the breadwinner and head of the household. The case of a wife who taught was quite different: her acceptance of a transfer tended to depend on the decision of the husband and his occupation; in this context, marriage could mean that a young woman would abandon her career or not pursue more attractive job alternatives in order to stay where her husband lived and worked.

The connection between marital status and the labor market was further complicated by the distribution of principals' positions. In cities, married women and widows comprised 16.1 percent of teachers and 37.5 percent of principals; in towns, the respective percentages were 25.8 percent and 34.3 percent, and in the country-side, 32.4 percent and 45.5 percent. The advantages favoring those who were not single can be explained, in theory, as evidence of a concerted effort on the part of school officials to give incentives to women teachers to keep them from leaving the teaching field, despite getting married.

The policy of rewarding married and widowed women who remained in the school system was, however, more than a strategy to retain personnel: at the same time, it reduced the impact the feminization of teaching had on gender and age hierarchies. The concern of male school authorities to place married women at the head of schools was an expedient means to limit the empowerment of single women, in whom converged a potentially explosive number of factors: public prestige, associated with teaching, an acceptable salary—for a woman—and the absence of control on the part of a husband. Single women, in such a context, made up 46 percent of the 916 teaching positions existing in 1904, but only 33.1 percent of the positions of principal.

A decisive transformation in the social position of a woman, even if she were married or widowed, was implied by her performing the duties of principal; and also, without a doubt, her representation in the public sphere was altered as well. The school hierarchy opened a space for women to exercise a formal and officially sanctioned authority.

Evidence of their quickly acquired power was their ability to place other family members in the school system: of the 95 schools headed by women, in 23 (24.2 percent) a relative also worked; the rate was only 14.1 percent in the schools headed by men (10 out of 71).

The relations of authority that were prevalent in schools were strongly conditioned by gender: of the 166 positions as principal, 92 were occupied by a woman who directed an all-female staff; 58 were held by a man who supervised the work of an all-male staff;[51] 13 posts were held by a man who supervised both male and female teachers,[52] and only three principalships (1.8 percent) were held by women who supervised men. The three schools in which this subversion of the public and official patriarchal order took place were located in the cities of Heredia and Puntarenas, and in the town of Esparza.

The spectrum of hierarchical relations, judging by the preceding data, were organized on the basis of a correspondence in gender between the principal and his or her staff: in 150 of the 166 principal's positions (90.1 percent), the lines of authority involved people of the same sex, although of different ages, a tendency that helps explain why married and widowed women—who were almost always older women—were found in positions of power. The separation of the male and female labor markets truly obeyed the interests of school authorities in improving their control of teaching personnel and perpetuating a discriminatory salary scale.

The scarce interconnection between these markets, and the partiality shown by educational authorities toward men, was the context within which, in the coming years, men teachers began to fight to defend the salary inequity. Women teachers, however, immediately affirmed that the wage discrimination was no longer acceptable, a position that reflected the degree to which schools were the site of a profound upheaval in gender relations. The school system, in effect, became the first component within the State in which women, once they had obtained and carried out positions of authority of an official, public nature began to subordinate male personnel in the occupational hierarchy.

CONCLUSION

The feminization of teaching in Costa Rica, which occurred early and rapidly in the Central American context, can be explained by the convergence of three processes: First, the increase in the demand for teachers caused by the reform of 1886, which the State met by hiring cheap female labor. Second, the growth of an urban economy, which offered young men with a certain degree of education job options that were more attractive than working as a primary school teacher. And third, an electoral dynamic which, on the one hand, channeled a growing proportion of State spending toward education, and on the other hand, favored the laboral insertion of women into public employment, above all, teaching.

The incidence of these structural factors was decisively strengthened by the role played by women, who quickly began to build a family tradition in teaching, to compete for principal's positions, and later, to challenge salary disparities and demand their right to vote. The limited integration of men and women teachers in primary education, during a time characterized by discrimination against women and an official preference for men, worked against the configuration of a professional identity that transcended gender. Kinship ties between women and men teachers were not sufficient to compensate for the separation between their spheres of work and their experiences, which was of such concern to school authorities. This division, which was already visible in the differentiated programs of study in use in the teacher training departments of the Colegio Superior de Señoritas and the Liceo de Costa Rica, promoted the acquisition of gender sociabilités and identities, especially among female personnel.

The foregoing process was facilitated by male "desertion" of the "army" of educators and by the "invasion" into this body by women. As this trend became more pronounced in the 1910s, salary discrimination and official favoritism toward men became increasingly unacceptable. The growing tension that resulted from this inequality was exacerbated during the administration of Luis Felipe González Flores (1914–1917), Minister of Public Education and defender of male intellectual superiority.[53] The dictatorship of the Tinoco brothers (1917–1919), which replaced the government of Alfredo González Flores (the brother of Luis Felipe), soon galvanized the opposition of educators, who effected a series of protests in June 1919 that hastened the fall of the dictatorship. The leading role played by women teachers and professors in those protests was, for a group of them, a decisive factor in taking up suffragism and feminism.[54] The struggle against the dictatorship helped to politicize the social and cultural discontent among women teachers that had been generated by the official discrimination against them.

The founding of the Costa Rican Feminist League (1923) was, as seen through this lens, an expression of suffragist grievances of a specific sector of women, the majority of whom were from the urban middle class; but at the same time, it implied the creation of a quasi-union, focused on gender issues. The first effective fight mobilized by this organization took place in 1924, when men teachers presented to Congress a law that proposed a salary raise exclusively for men. It was quickly and strongly challenged by the League. According to its president, the lawyer Ángela Acuña,

> [Y]es, at that time I stirred up the interest of women teachers across the country. I sent telegrams to all the women principals of all the schools in the Republic. I presented Congress with a protest petition signed by an enormous number of women teachers . . . In the Porfirio Brenes school there was a secret meeting of men teachers in order to demand a unilateral campaign. Women teachers joined with me to fight the decisive battle . . . Men lost the game.[55]

The union-like success of the League in 1924 stands in contrast to their defeat a year later when, despite their efforts, universal suffrage was thrown out by Congress. The weight of the patriarchal ideology among the legislators is frequently identified as the cause of this defeat; but other factors are worth considering: the opposition of political parties to doubling the number of voters—with the resulting uncertainty—on the eve of elections for members of Congress in 1925; the lack of a popular base that supported the vote for women;[56] and the fear among politicians that the granting of such a right would, in the long run, open the market of elected offices (especially in Congress and municipalities) to competition by women.

The challenge was not long in coming. In 1925, it was clear that women, and particularly those from cities and from upper- and middle-class backgrounds, were capable of organizing and mobilizing, even under extreme conditions such as those in June 1919; that they had at their disposal specific strategies—family ties—to consolidate their position within the occupations that they had "invaded," for example, teaching; that they could rely on a quasi-union, the Feminist League, which had been fairly successful; that they had learned to make their civic virtues visible in national life through means of print culture;[57] and if that were not enough, that they were prepared to knock on the doors of Congress to demand their voting rights.

The complaint made in 1913 by Rómulo Tovar about the grave deficiencies among staff working in schools, had an empirical basis in their low rates of certification: 13.5 percent in 1911. The increase in the rates of certified teachers (72.8 percent in 1926)[58] did not imply, however, that the criticisms of their preparations had ended. In 1925, the Minister of Public Instruction, Napoleón Quesada, attributed the failings of education to the imbalance between the sexes:

> In my view, the students in the first and second grades, which are natural extensions of the home, need affection, patience, and a mother's love, and should be taught by women teachers who put a woman's heart into their work. But educational work performed by women in the third, fourth, fifth and sixth grades distorts the character and compromises the future of young boys. . . . I detest the intimidated, sentimental, submissive character among youths who graduate [from] schools where they have heard nothing but a woman's voice, and have resolved to do nothing more than what was suggested by a woman.[59]

According to the highest educational authorities, the predominance of women in primary education posed a very grave danger: the emasculation of young boys and men, which foreboded a country of effeminate men! The persistent and sharp criticisms over the preparation of teaching staff in Costa Rica, despite the increased percentage of certified teachers, demonstrate that such complaints—without excluding the existing deficiencies in teaching—were beyond all else a tool that was useful for men to challenge the visibility of women and femininity in the public sphere. The fact that this last process would depend on the feminization of teaching, which was basic to stimulating the formation of feminist intellectual circles, helps explain why the work of women teachers was so discredited.

The constant criticism of the work performed by women educators constituted, from this point of view, an obvious effort to label primary education as unprofessional, since it was an occupation dominated by women.[60] This image in turn contributed to the desertion by men of teaching. On the other hand, the construction of a systematic doubt concerning women's professionalism was key in allowing male political and intellectual circles to maintain control of the best job, promotion, and supervisory opportunities within the public sector, which was for these groups a strategic wellspring of resources and employment.

Notes

1. Education in Costa Rica was under municipal control—as stipulated by the 1812 Constitution of Cádiz—until 1847, when in the Constitution of that same year the Ministries of the Interior, Public Education, War, and the Navy were created. The municipalities, however, retained a strong influence over teaching until the reforms of 1886. Ileana Muñoz, *Educación municipal en Costa Rica: 1821–1882* (San José: Editorial de la Universidad de Costa Rica, 2002).

2. Rómulo Tovar, *Don Mauro Fernández y el problema escolar costarricense* (San José: Imprenta Alsina, 1913), p. 67.

3. Costa Rica, *Memoria de Instrucción Pública* (San José: Tipografía Nacional, 1904), p. 129. See also Mario Matarrita, "El desarrollo de la educación primaria en Costa Rica," in Carmen Lila- Gómez et al., *Las instituciones costarricenses del siglo XX* (San José: Editorial Costa Rica, 1986), p. 150.

4. Costa Rica, *Organización del personal docente de las escuelas primarias* (San José: Tipografía Nacional, 1904).

5. Hermógenes Hernández, *Costa Rica: evolución territorial y principales censos de población 1502–1984* (San José: Editorial Universidad Estatal a Distancia, 1985), p. 173.

6. Iván Molina, "Ciclo electoral y políticas públicas en Costa Rica (1890–1948)," *Revista Mexicana de Sociología* (México, LXIII (3) July–September 2001): 74, 80.

7. Ibid., 74, 81.

8. The districts were: San Isidro (1910), San Juan (1914), San Vicente (1914), Alajuelita (1909), Curridabat (1929), San Pedro (1915), Sabanilla and Santa Ana (1907), in San José; Los Ángeles, in Cartago; and San Isidro (1905), San Pablo (1961), San Joaquín (1907) and San Antonio (1915) in Heredia. The date between parentheses indicates the year in which it became a canton. The canton of San Mateo in Alajuela is listed in the province of Puntarenas in the 1904 register.

9. Costa Rica, *Memoria de Instrucción Pública*, p. 156.

10. Mario Samper, *Generations of Settlers. Rural Households and Markets on the Costa Rican Frontier, 1850–1935* (Boulder, CO: Westview Press, 1990).

11. For three exceptions, see Marcia Apuy, "Desarrollo de la educación femenina en Costa Rica (1889–1949)," in Elías Zeledón (ed.), *Surcos de lucha. Libro biográfico, histórico y gráfico de la mujer costarricense* (Heredia: Instituto de Estudios de la Mujer, 1997), pp. 311–318. Virginia Mora, "Rompiendo mitos y forjando historia. Mujeres urbanas y relaciones de género en el San José de los años veinte" (Tesis de Maestría en Historia, Universidad de Costa Rica, 1998), pp. 237–238. Iván Molina and Steven Palmer, *Educando a Costa Rica. Alfabetización popular, formación docente y género (1880–1950)* (San José: Plumsock Mesoamerican Studies y Editorial Porvenir, 2000), pp. 81–82.

12. Costa Rica, *Censo de la República de Costa Rica* (San José: Tipografía Nacional, 1885–1886), pp. 86, 88. Women represented 37.6 percent (44 of 117) of teachers counted by the 1864 census. Costa Rica, *Censo general de la República de Costa Rica (27 de noviembre de 1864)* (San José: Imprenta Nacional, 1868), p. 96. The percentage of women, who worked as teachers in Costa Rica in the 1880s, was similar to the percentage in Spain in the same era (44.4 percent). See Sonsoles San Román, *Las primeras maestras. Los orígenes del*

proceso de feminización docente en España (Barcelona: Editorial Ariel, 1998), p. 219.

13. Gustavo Niederlein, *The Republic of Costa Rica* (Philadelphia: The Philadelphia Commercial Museum, 1898), p. 78.

14. Ibid., 78. Terencio Sierra, "Mensaje," *El Pabellón de Honduras. Semanario semi-oficial,* Tegucigalpa (January 4, 1902): 1. Bradford E. Burns, "The Intellectual Infrastructure of Modernization in El Salvador, 1870–1900," *The Americas,* XLI, 3 (January 1985): 79. Apuy, "Desarrollo de la educación femenina en Costa Rica," p. 314. Guatemala, *Censo general de la República de Guatemala de 1921* (Guatemala: Tipografía Nacional, 1924). *Diario Nuevo* (San Salvador: January 18, 1934), p. 3. Cameron D. Ebaugh, "Education in Nicaragua," *Federal Security Agency Bulletin,* 6 (1947): 14. I thank Blanca Ileana Ordóñez and Óscar Peláez for the data on Guatemala.

15. Costa Rica, *Memoria de Instrucción,* p. 118.

16. Apuy, "Desarrollo de la educación feminina en Costa Rica," p. 313. For a comparison with the French case, see Jo Burr Margadant, *Madame le Professeur. Women Educators in the Third Republic* (Princeton, NJ: Princeton University Press, 1990), pp. 21–22. The number of male teachers in Costa Rica increased from 252 in 1892 to 322 in 1900, and fluctuated between a minimum of 313 and a maximum of 360 individuals in the 1920s.

17. Carlos Luis Fallas, *El movimiento obrero en Costa Rica 1830–1902* (San José: Editorial Universidad Estatal a Distancia, 1983), p. 339. The Costa Rican currency changed from the peso to the colón in 1900.

18. Mario Oliva, *Artesanos y obreros costarricenses 1880–1914* (San José: Editorial Costa Rica, 1985), pp. 59–61. Fallas, *El movimiento obrero en Costa Rica,* pp. 337–338.

19. Víctor Hugo Acuña and Iván Molina, "Base de datos del Censo Municipal de San José de 1904" (San José: Centro de Investigaciones Históricas de América Central, 1992–1997).

20. Fallas, *El movimiento obrero en Costa Rica,* p. 339.

21. Eugenia Rodríguez, " 'Que la mujer brille y se enaltezca por sus virtudes,' Selección de documentos sobre las tipógrafas josefinas (1903–1912)," *Revista de Historia,* San José, 33 (January–June, 1996): 145.

22. The situation in Great Britain was similar. See Alison Oram, *Women Teachers and Feminist Politics 1900–1939* (Manchester: Manchester University Press, 1996), pp. 24–25. Dina M. Copelman, *London's Women Teachers. Gender, Class and Feminism 1870–1930* (London: Routledge, 1996), pp. 75–79.

23. The University of Santo Tomás, founded in 1843, was closed in 1888, within the context of the educational reform of 1886. The only fields of study that remained open were in law and pharmacy; those who wanted to study medicine had to do so in another country. The nation

lacked a university until 1940, when the University of Costa Rica was established. See Paulino González, *La Universidad de Santo Tomás* (San José: Editorial de la Universidad de Costa Rica, 1989).

24. Amitai Etzione (ed.), *The Semi-professions and their Organization: Teachers, Nurses, Social Workers* (New York: Free Press, 1969); and Dan Lortie, *Schoolteacher: A Sociological Study* (Chicago, IL: University of Chicago Press, 1975).

25. Costa Rica, *Memoria de Instrucción*, p. 117.

26. Molina and Palmer, *Educando a Costa Rica*, pp. 81–87. In Costa Rica equal access to teacher training schools for men and women has existed since these schools were opened to the present day. No law has ever privileged or discriminated against any student's admission on the basis of sex.

27. Margarita Silva, "La educación de la mujer en Costa Rica durante el siglo XIX," *Revista de Historia*, San José, 20 (July–December 1989): 71. In 1870 a new teacher training school was opened in San José, but it did not register enough students and was closed in 1871. Juan Rafael Quesada, "La educación en Costa Rica: 1821–1914," in Ana María Botey (ed.), *Costa Rica, Estado, economía, sociedad y cultura desde las sociedades autóctonas hasta 1914* (San José: Editorial de la Universidad de Costa Rica, 1999), pp. 381–382.

28. Ástrid Fischel, *El uso ingenioso de la ideología en Costa Rica* (San José: Editorial Universidad Estatal a Distancia, 1992), pp. 97–98. The teacher training school—which is still little investigated from a social and cultural perspective—was one of the bases of the Universidad Nacional, which opened in 1973.

29. Molina and Palmer, *Educando a Costa Rica*, pp. 81–92. The teacher training section of the Colegio Superior de Señoritas began operating again between 1918 and 1923.

30. For a comparison to the United States, see Myra H. Strober and Audri Gordon Lanford, "The Feminization of Public School Teaching: Cross-Sectional Analysis, 1850–1880," *Signs* 11, 2 (1986): 212–235.

31. Ólger González, "Análisis histórico de la evolución burocrática en las distintas carteras gubernamentales de Costa Rica, 1871–1919" (Tesis de Licenciatura en Historia, Universidad de Costa Rica, 1980), Table 3.

32. Iván Molina, *El que quiera divertirse. Libros y sociedad en Costa Rica (1750–1914)* (San José: Editorial de la Universidad de Costa Rica y Editorial Universidad Nacional, 1995), p. 134.

33. Costa Rica, "Población de la República de Costa Rica al 31 de diciembre de 1904," *Ley de Elecciones*, 2ª, ed. (San José: Tipografía Nacional, 1905), pp. 39–49. This data appears to reflect an overcount of 8.5 percent.

34. Information on salaries is found in: Costa Rica, *Memoria de Instrucción*, p. 64.

35. An interesting study on this subject in France is found in Linda L. Clark, "Bringing Feminine Qualities into the Public Sphere. The Third Republic's Appointment of Women Inspectors," in Elinor A. Accampo, Rachel G. Fuchs, and Mary Liynn Stewart (eds.), *Gender and the Politics of Social Reform in France, 1870–1914* (Baltimore, MD: Johns Hopkins University Press, 1995), pp. 128–156.

36. Costa Rica, *Memoria de Instrucción*, pp. 51, 54, 55, 101. Information about how much Dávila earned for his work in the Escuela de Comercio is not available.

37. Costa Rica, *Memoria de Instrucción*, pp. 54–55.

38. Apuy, "Dessarolo de la educación feminina en Costa Rica," p. 318. On gender inequalities in the academic world, see the interesting contributions of Sandra Acker with respect to Great Britain. Sandra Acker, "New Perspectives on an Old Problem: The Position of Women Academics in British Higher Education," *Higher Education* 24, 1 (July 1992): 57–75; Sandra Acker and Grace Feuerverger, "Doing Good and Feeling Bad: The Work of Women University Teachers," *Cambridge Journal of Education*, 26, 3 (November 1996): 401–422.

39. Costa Rica, *Memoria de Instrucción*, pp. 64–67; Costa Rica, *Organización del personal docente*, pp. 10, 31.

40. The available information does not show that laws had been passed in Costa Rica prohibiting a married couple from working in the same school. When Luis Felipe González-Flores was Minister of Education during the presidency of his brother Alfredo (1914–1917), he succeeded at excluding married women teachers from teaching, a policy that increased discontent against the government. Ástrid Fischel, "Estado liberal y discriminación sexista en Costa Rica," *Revista de Ciencias Sociales*, San José, 65 (September 1994): 29–30.

41. These cases seem to correspond to the model of the maternal teacher. See in this respect San Román, *La primeras maestras*, pp. 29–33. About the conception held by Latin American liberals of women as natural teachers, see Molina and Palmer, *Educando a Costa Rica*, pp. 60–61.

42. Oresta López documents a similar process in rural areas in the Valle del Mezquital in México. Oresta López, *Alfabeto y enseñanzas domésticas. El arte de ser maestra rural en el Valle del Mezquital* (México: CIESAS, 2001), pp. 164–167.

43. Oconitrillo, Eduardo, *Los Tinoco (1917–1919)*, 3rd. edition (San José: Editorial Costa Rica, 1991), pp. 159–172.

44. The person who held the position of principal and teacher in four of those 74 schools had responsibilities for the boys' school as well as the girls' school.

45. The number of principals (75) didn't coincide with the number of schools (81) because in the towns there were five people who each

directed two schools, and one school without the position of a principal.

46. The average decreases to 9 and 4.1 positions per principal in cities and towns, respectively, when teachers of special subjects are excluded.

47. Fallas, *El movimiento obrero en Costa Rica*, p. 339. Costa Rica, *Memoria de Instrucción*, pp. 62, 64. The salary of the woman principal of the girls' school in Grecia was increased to 60 colones a month in August, 1903.

48. Costa Rica, *Censo de población de Costa Rica 11 de mayo de 1927* (San José: Dirección General de Estadística y Censos, 1960), p. 23.

49. Acuña and Molina, "Base de datos del Censo Municipal de San José de 1994."

50. Costa Rica, *Memoria de Instrucción*, p. 201.

51. The 58 cases included a position for principal and teacher in a school where only one teacher was in charge.

52. The 13 cases break down as follows: there were five in which the hierarchical superior/principal supervised female staff, and eight in which the principal supervised teachers of both sexes.

53. Fischel, "Estado liberal y discriminación," pp. 29–30.

54. Oconitrillo, *Los Tinoco (1917–1919)*, pp. 159–172.

55. Ángela Acuña de Chacón, *La mujer costarricense a través de cuatro siglos*, II (San José: Imprenta Nacional, 1969), p. 357. Macarena Barahona, *Las sufragistas de Costa Rica* (San José: Editorial de la Universidad de Costa Rica, 1994), p. 84. The unrest in 1924 was repeated in 1928, when a salary increase for men teachers was again proposed and once again, failed.

56. Iván Molina and Fabrice Lehoucq, *Urnas de lo inesperado. Fraude electoral y lucha política en Costa Rica (1901–1948)* (San José: Editorial de la Universidad de Costa Rica, 1999), pp. 101–104.

57. Eugenia Rodríguez, "La redefinición de los discursos sobre la familia y el género en Costa Rica (1890–1930)," *População e Familia*, 2, 2 (São Paulo, July–December 1999): 147–182.

58. Matarrita, *El desarrollo de la educación primaria en Costa Rica*, p. 150.

59. Fischel, "Estado liberal y discriminación," pp. 34–35.

60. Primary school teaching would thereby be a semi-profession. See in this regard: Etzione, *The Semi-professions and their Organization* and Lortie, *Schoolteacher: A Sociological Study*.

CHAPTER 8

Society and Curriculum in the Feminization of the Teaching Profession in the Dominican Republic, 1860–1935

Juan Alfonseca Giner de los Ríos

INTRODUCTION

This chapter develops certain ideas about the importance that should be attributed to society and curriculum during the feminization of the teaching profession in the Dominican Republic from 1860 to 1935. In this period, the Dominican national school lost the distinctly masculine characteristics that had defined it during the nineteenth century. It reconfigured itself as a space in which women became the predominant actors within the pedagogical practice. As was the case in other less developed societies in which mandatory public education was a late arrival, the preponderance of women occurred suddenly, and was realized in a much shorter period than is covered by this study. It took less than four decades for the gender distribution of teaching positions to be reversed.[1]

What importance did society and curriculum have during the massive and growing incorporation of women into the teaching profession in the Dominican Republic? What character did each have as a determinant factor in the process of feminization of teaching? As a *curricular determinant* in the process of feminization I have assumed that codified *prescriptions* exist in the design of educational models that systematically produce feminization. As a *social determinant*

I assume the plausibility of historical processes of a demographic and cultural nature that induced feminization, such as migrations, the process of work, and social representations that have arisen around education. I have not imagined, *a priori*, a definite conceptual split between the one and the other. What I define as social in this chapter (e.g., the use of women teachers in schools for minority communities of ethnic migrants) is saturated with curriculum: it is a cultural behavior presided over by curriculum, as is the common notion that a woman's physical weakness impedes her ability to educate male students. In proposing an analysis that differentiates between curriculum and society and the importance that should be ascribed to each during the process of feminization, I will attempt to distinguish between the curricular actions of the Dominican State and its intellectuals from those actions that can be attributed to society, which were a historical given. In other words, I ask what part of the feminizing process was due to foundational actions of the State and what part is due to the historical configuration of Dominican society. The analytical tension I wish to establish is between a social genesis and an institutional genesis.[2]

The treatment of this curricular, discursive and pedagogical dimension is, for different reasons, somewhat unfocused. Foundational texts on gender distinctions, social practices, and social discourses that make representations about the suitability of the woman teacher are seldom found in the educational discourse or in the programmatic action that the Dominican State proposed to implement.

If one reviews the pedagogical writings that serve as foundational texts for public education in the Dominican Republic, there is little pertaining to female participation in teaching. In this sense, the intellectual production of these men of letters[3] scarcely contemplated the development of a general program of educational reforms that would facilitate the modernization of a deeply traditional agrarian society whose political structures were highly unstable. A fragile institutional framework had emerged from a recent colonial past characterized by decadence, isolation, cultural domination (the conquest of the Spanish colony of Santo Domingo by black Haiti), and the absence of a centralizing national impulse.

No significant curricular concern regarding feminization can be detected in the pedagogical and organizational policies that favored public education. Reforms that had been influenced by the rationalist pedagogy of Eugenio María de Hostos during his stay in the country—reforms which advocated the foundation of Normal Schools—produced few results with respect to differentiated education until the turn of the century. Soon thereafter, the country would be paralyzed by a crisis that ultimately provoked the first military intervention by the United States in the Dominican Republic in 1916.

Before the U.S. Military Government of Occupation was installed, the Public Education Code of 1914[4] formalized the pedagogical design that had been in place in the country since the Restoration movement of the 1860s. However, the Code included no definitions of specific educational devices such as primary education, separate schools for boys and girls, or nursery schools. This stands in contrast to the processes of codification in other countries, in which the character of the teacher and the degree of female participation in teaching were also clearly defined.[5]

In 1916, the Occupation Government, interested in projecting a favorable image of the military government, implemented mandatory public education and undertook a significant expansion of schools throughout the country. Although their direction and administration would be left in Dominican hands, the goal was to extend educational services in a short period of time. The cultural politics of the rudimentary schools[6] in the years 1916–1924 revolved around public health, the formation of local associations, the development of school gardens and the creation of a national imaginary. Coeducation, which was adopted by primary schools in those years, did not generate large debates. Its limited introduction suggests that the educational *ethos* of the communities intervened, and in the mixed schools ways of separating the sexes were reinstituted.[7]

A new phase of political instability following the withdrawal of U.S. occupation forces prevented the National Service of Public Instruction[8] from developing the Dominican educational system in a pedagogical and programmatic way. The central educational policies in the period from 1924 to 1930 maintained the organization of the prior school system and focused more on reversing the effects of the economic crisis in 1921, which was caused by the fall in international sugar prices. Schools in rural areas had been especially hard hit, and a large number of them had been closed.

In 1930, the ascent to power of the man who would later become dictator, Rafael Leónidas Trujillo, brought new changes to the educational system, which very quickly organized itself under the apparatus of ideological and social control in the service of the *Benefactor de la Patria*. During this period in the early 1930s, it becomes possible to trace the emergence of an imagery, deliberately animated by those in power, in which the feminine is the moral bastion of the *Patria Nueva*—the feminine and all that corresponds to women: the mother, the woman teacher, the female nurse, the wife, all of whom were the carriers and builders of a new ethical order during the *Trujillo Era*.

El trujillismo stimulated a clear gender image in the Dominican educational system. Pedagogy and primary school education, special education, and arts and crafts were incorporated into Dominican education in the early thirties, becoming emblematic areas of curricular and institutional action for the regime. Teaching also played an important role in the construction of the *Trujillo Era*: teachers were guardians of public peace in their domain, involuntary members of the *Partido Dominicano*, champions of solidarity with the regime and, as one of its central images of good citizenship, instruments of the propaganda machine. The *Benefactor* himself was the First Teacher of the Republic, just as his mother, *la Ilustre Matrona*, gave loving instruction in morality to Dominican society. In the theater of ideas about progress, teachers were the builders of an exalted nation. As agents of modernization for the campesinos, and intermediaries through whom arts and crafts were spread, teachers were also the representatives of *la dominicanidad* in the *Escuelas Fronterizas* near the Haitian frontier, where they implemented the politics of spiritual "Dominicanization."

Although it was not until 1930 that a cultural politics of gender in Dominican education emerged, during the previous seven decades the national school system maintained practices and organizational patterns inflected by gender whose impact on the process of feminization still need to be investigated. To what extent was the design of the curriculum in those years a result of these patterns or, in terms of its educational practices and discourse, to the legacy of the colonial past? To what extent did they also derive from internal cultural processes that were present in Dominican society, processes that were not dependent on State interventions but instead derived from cultural behaviors brought by immigrants to the country, from demographic changes in different regions, and from the intersection of political power and education? In other words, what are the factors underlying the feminization that the Dominican teaching field experienced during the period in question? In the pages that follow I will attempt to illustrate several aspects of the phenomenon.

This chapter is the result of an investigation of documentary sources generated by the National Service of Public Instruction between 1915 and 1935, a period that opens with this agency's emergence as a centralized office in charge of implementing mandatory popular education. I cover the development of its policies until the end of the first five-year-period of the Trujillo dictatorship. This is a preliminary study of the early stages of the school system in the Dominican Republic, and utilizes the limited number of historical studies on the Dominican national school.[9] It is intended to serve as a precursor to other investigations into the emergence and establishment of processes such as the *feminization of teaching*

Many of the ideas set forth herein derive from a systematic examination of payroll records for teachers at a local level. This is an excellent strategy to reconstruct the feminization of teaching because it overcomes the normal limitations of national statistics. Also it reveals the regional differences that characterize feminization, which was a process that developed unevenly and was impacted significantly by local contexts and cultures.

In addition, the individualized information that these payroll records contain make it possible to establish possible family relations and to determine factors relating to the job continuity and mobility of teachers.

The Participation of Women in Dominican Education: Main Characteristics and Trends, 1860–1915

In the years before 1860, the effort to construct a juridical and institutional code of laws that normalized and expanded educational practices had advanced very little. Likewise, little progress had been made in securing public order in the nation, as only fifteen years had passed since the new republic freed itself from Haitian domination.

The first law, the Public Instruction Law of May, 1845,[10] decreed the creation of free and public primary schools by the local governments in each *común*,[11] reserving for the State a rather loosely defined margin of curricular control. As Morrison points out, this law did little to advance the growth of public schools, for of the 32 primary schools that the counties were to fund, only nine had been established by 1850,[12] a situation that points to limited revenues and a lack of interest among the political powers and local societies in promoting schools. This state of affairs would continue for quite some time, despite measures that provided for freedom of education and promoted the establishment of private schools.[13]

Education was a scarce resource in Dominican society before 1880. Pedagogical practices of a diverse nature existed. The economic and political elite instructed its children in private schools, most of them denominational, which were organized under the traditional pattern of gender division: girls' schools and boys' schools. Below the cluster of elite schools, the urban middle class sent its children to a network of mostly domestic spaces, where they were instructed in one skill (reading or writing or arithmetic, and sometimes all three) without a graded curriculum. It is difficult to determine the scope of this network. For example, in 1860 Santo Domingo, the capital, had 35 schools whose name was identical to the instructor's, indicating that small and separate groups of children and adolescents were educated in the teacher's home. Of the 35 individuals who figure on that list of schools, 71 percent (25 individuals) were married women, widows, or single women who taught boys and girls. Only ten men teachers ran this type of establishment for all-male students.[14]

There were also schools for craftsmen and workers, sometimes run by guilds and fraternities, others born out of enlightened initiatives of the great men of the Republic. With respect to the tobacco craft, Hoetink describes such practices as readers who would entertain tobacco workers and lead discussions during work shifts.[15]

In truth, statistics and descriptions from that era express only part of the educational movement that was advanced by society. For example, they reveal few traces of the schools created by Protestants in order to educate the liberated American slaves who came to the country during the years of the Haitian domination. These schools comprised a network that went beyond their most visible institutions in the cities, such as the famous Wesleyan school in Puerto Plata where Gregorio Luperón, the hero of the Restoration, received an education. Additional schools were located within small rural congregations, especially on the Samaná peninsula.[16] Besides the network of Protestant schools, a far from negligible number of home schools specialized in the education of ethnic minorities who had migrated from other parts of the Caribbean, such as Haiti, Curaçao, or the "*cocolos*" from English-speaking islands such as St. Kitts, Nevis, the Virgin Islands, the Turks Islands, and others.

The impact of this transplanted school culture on local school practices with respect to coeducation and expectations about the image and gender of the teacher, deserve a closer study. The National Service of Public Instruction was slow to Dominicanize the so-called "English" schools established by the *Yankee*[17] ex-slaves in Samaná. In those schools English was the language of instruction; the Spanish language, national subjects, and the pragmatic objectives of pedagogical work—the national Dominican curriculum, in short—were left to the side in order to leave room for the transmission of the Protestant faith and allow cultural contact with the nation of origin. As a pedagogical space, the "English" schools followed an organizational pattern that associated them with the church and the *Miss*, who was at times the preacher's daughter. A clearly feminized pattern existed that impacted the history of schools on the peninsula, thus illustrating the social origins of the global phenomenon of feminization.

One of the first full-scale school censuses that was done in the country was published in 1882 in the Report of the Secretary of State of Justice, Development, and Public Instruction.[18] According to this census, 127 centers of education existed in the national territory, of which the majority were only at the elementary school level. Although the 1881–1882 census is incomplete, it provides a sample of the gender division within the job market for Dominican teachers at that time, showing that a third of all teaching positions were held by women who work mainly in elementary education.

The census show that Dominican schools were predominantly male. It reveals that educational relations were organized according to traditional gender models: girls' schools were taught by women; boys' schools were taught by men; mixed schools were for the most part taught by women; and high status knowledge was controlled by men.

Although generalized in nature, the pattern exhibited variations (six girls' schools run by men teachers; six schools for both sexes also run by men, in addition to men's greater presence in higher grades for girls), which stands in contrast to the much less flexible pattern in boys' education: in all-male schools there is no female presence. Boys are educated by men; men educate men. Young Men's Schools are, like evening schools for workers and craftsmen, invariably staffed by men. This was not based on any curricular principle, at least at the elementary school level, and may have followed the common dictum that masculinity could not be controlled by women, as a protest made many years later confirms. Residents in the town of Estancia del Yaque, County of Santiago, complained to President Horacio Vásquez about Idalia Tavarez, a recently hired principal, in the following terms:

> Women teachers . . . haven't produced any results and the school has been run very badly. These women aren't given any respect by the boys, who are very bad in this place. The worst was when Mrs. Amparo Marmolejo ran the school, because . . . that teacher couldn't or didn't want to draw the line . . . one boy beat another one with a stick and killed him . . . it all happened because of the bad habits the boys had acquired with these weak women teachers.[19]

The pattern of gender distribution shown in the 1881–1882 census responded to a pedagogical *ethos* that was representative of social attitudes concerning education. To a lesser extent, it is the expression of the State's interest in regulating school relations, since the sector made up of schools founded through private initiative (communal schools, private schools, schools run by trade organizations, etc.) was in the majority, thus demonstrating the importance of traditions and popular imaginaries with deeper historical roots.

In large measure, women's insertion into elementary school teaching was characterized by the development of traditional socialization tasks related to sewing and religion. Although at that time a study plan for elementary schools existed that also listed reading, writing, grammar, and comportment, the instruction given to the girls by their women teachers tended to be in those two areas alone. Regarding girls' education, Pedro F. Bonó—an intellectual with reformist ideas and the ex-Minister of Justice and Instruction—pointed out in a critical article published in *El Eco del Pueblo* in 1884:

> [W]hat these girls are taught doesn't fulfill the purpose and mission assigned on Earth for the fairer sex. The lower classes have been deprived of the national traditions of basic sewing, of making their own shirts and socks, of the Christian catechism and domestic work. These have been replaced in great part by lacemaking, oratorio, and tapestry making. The Government shouldn't defend such misguided aspirations of poor parents, of all our working classes, nor should they be encouraged. This is not the path that will elevate the girls as their parents wish; they are not likely to send their daughters to the Legislative Chambers, nor to the Gothic Court in Flanders, nor to teach at the Gobelin tapestry workshop; they are being educated for marriage, they are being instructed to be the mother of a family, but I ask: can these poor lace and tapestry makers without a dowry, without a trousseau, be the happy wives of the hovel-dwellers who await them, with infinitesimal, arbitrary, and fleeting wages?[20]

Bonó's commentary shows us the existence of a national tradition of gender-based education. He was troubled by impractical innovations in this new curriculum that abandoned instruction that was appropriate and useful for working-class girls. Yet this new curriculum reproduced the same pattern of differentiated instruction. Bonó also reveals a position typical of enlightened and reformist intellectuals with respect to the education of the *fair sex*, especially the *fair sex* of the working and rural class: her mission was to be productive, so that the class would survive and develop: to embroider goods with a practical use, to manage the home, and to behave in a Christian manner. From the social reformist perspective of the Valle del Cibao intelligentsia, who fought for the development of a self-sustaining market, these traditions in women's education were preferable because they gave the working class tools for its reproduction[21] that were superior to those provided by the new curricular guidelines and their illusion of social ascent. Furthermore, Bonó's commentary provides historiographic evidence of a differentiated education for boys and girls, one that determined the functions, images, and conditions for women's insertion into the teaching field.

It is difficult to determine the personal or everyday characteristics of these women on the basis of the limited information available. An analysis of an 1860 payroll for teachers who worked in their homes allows us to affirm that close to a third of the women listed therein were or had been married: six of them added *Viuda* [widow] to their signatures[22] and 12 signed their last names with "*de*," suggesting that they were married.[23] To conclude that the remaining women teachers were single because of the absence of a "*de*" or a "*Viuda*" in their listed names is risky because marital ties could be concealed, as in the case of teachers with an Anglo-Saxon origin, for whom it was customary to adopt their husband's surname. This practice was also followed to some degree by Dominican women.

In terms of jobs, the best conditions were reserved for men. In the county of Santo Domingo, for example, one male primary school principal received a salary that was approximately 40 percent higher than the female principal of the Higher Institute for Girls who, unlike her male counterpart, was also given teaching duties. With less severe salary inequalities, this situation is reproduced in other cities in the interior, such as Azua, La Vega, and El Seybo.[24]

A comparison of the teacher payrolls in the city of Santo Domingo for the years 1860 and 1881 gives rise to certain conjectures. Only four of the twenty-five women who were teachers in 1860 were still teaching twenty years later, which indicates that job stability for teachers was still not as pronounced as it would later become as part of a worldwide trend. More significantly, none of the surnames of the twenty-one remaining teachers reappear in 1881, which suggests there was no tradition of teaching in these families.

Further speculations can be made despite the lack of sources. Of the four surnames that are repeated, two are for the *Señoritas Bobadilla—Eugenia and Federica—*who operated schools in their homes in 1869 and who in 1881 continued to work in schools for girls run by the County of Santo Domingo. In the work *Familias Dominicanas* compiled by Carlos Larrazabal Blanco, a *Eugenia Bobadilla* is listed, the daughter of Miguel Bobadilla and Manuela

Viera, residents of the San Carlos district outside the capital. The *Eugenia* recorded by Larrazabal died, without children, in 1892 at age 77. If this is the Señorita Eugenia Bobadilla who appears on the lists in 1860 and 1881, then it may well be that she was a single teacher, around 50 years old in 1860.[25] A plausible conjecture is that the presence of widows, unmarried women—*señoritas* who were along in years—was consistent with the cultural model requiring that girls be given an education that was grounded in morals.

Immigrant teachers from other areas were also at work in the Dominican Republic during these years. For example, it is well known that members of the political-military elite of the War of Restoration were educated in schools run by Moravians, both inside and outside the country. Ulises Hereaux studied in the "English" school of Mr. Thauller, in Puerto Plata. Hereaux's daughters, Dilia, Herminia, and Casimira, studied in the *Welgelegen* boarding school run by Franciscans in Roosendaal, on the neighboring island of Curacao, where daughters of other Army members studied as well. A seasonal wave of migrants came from this island for the purpose of teaching the Dominican elite.[26]

To the Wesleyan Methodists in the cities of Puerto Plata, Santiago, and Santo Domingo—teachers who had come directly from Wales and Scotland (such as Thauller, Mears, and Gooding)—can be added the names of other teachers in the denominational school network, such as the Ashtons from Saint Thomas and the Virgin Islands, the Vanderhorsts (presumably from the United States), the Beers, and other such.[27]

This stream of teachers is distinct from another that specialized in educating Francophone migrants. Around 1915 it was common to encounter in border cities, especially in merchant ports such as Montecristi, San Pedro de Macorís, Puerto Plata and Samaná, schools in French Creole.

Men and women teachers in "English" schools; "patois" schools; teachers from Cuba and Puerto Rico fleeing wars of independence; Mexican and Spanish nuns—all brought to the incipient Dominican school culture a diverse set of practices and cultural horizons regarding education (and the role gender would play in it) from their native schools.

The 1914 Code of Public Education introduced a series of principles that not only outlined a gender model for the national school system, but also set a firm basis for the nationalization of the curriculum, which would fall upon the U.S. Occupational Government to implement after 1916. It was at the time that the national curriculum was imposed that gender practices in the schools founded by immigrants were challenged and confrontations occurred.

Such was the case with the "English" schools on the Samaná peninsula, where one of the first and longest acts of cultural resistance arose against the national curriculum and its mechanisms of control regarding such aspects as gender relations in education. The terms of this cultural confrontation can be seen in the request made by the Protestants in Samaná to the U.S. Military Governor during his visit to the peninsula:

1. That in the Sections where the house provided for the school is not the property of the government, but is rented, and therefore does not belong to the Popular Boards of Education but to the Church, religious services of these denominations shall be permitted during non-school hours, for preaching, performing marriages, baptisms, etc., and that [all members of] these denominations have the right to attend without distinction.

2. That instead of dividing the time . . . between boys and girls, three hours in the morning and three hours in the afternoon, all boys and girls shall attend in the morning and afternoon.[28]

To this request the Protestants of Samaná later added an additional one concerning the preferred gender of their teachers. As Inspector Zúñiga reported shortly thereafter:

The Board of Education of Clará, Samaná, requests as principal for School 9 in that district a woman teacher who is Protestant and can speak English, because almost all who live there are English-speaking Protestants.[29]

The Emergence of a Feminized System in Dominican Public Schools, 1916–1930

In April 1918, Rear Admiral H.S. Knapp, the U.S. Military Governor in Santo Domingo, decreed Executive Order No. 145, which turned into law the recommendations made by the Commission of Dominican Education concerning public education.

Although the Organic Law of Public Education substantially reiterated the 1914 Public Education Code, it simplified the curricular structure of the school system, suppressing (to the degree it made no reference to it) the different spaces and separate study programs for boys and girls.

Most importantly, the Executive Order caused a significant expansion of free, mandatory public education, especially in rural zones where, in the years 1918–1922, an extended process of cultural encounter with campesino societies took place, which for the most part had lacked any previous experience with schools.

Paralleling the growth in primary education was an increase in educational employment, since the Occupation government's initiative to expand mandatory education generated a sudden and disproportionate demand for teachers. The teachers hired to work in the still-undeveloped educational system displayed a diverse profile: individuals with little education, individuals from other school systems, graduates of teacher training programs, and so on.

We lack detailed sources that would allow us to determine how the national school system grew region by region between 1881 and 1916. Most likely, it was urban education (in the capitals of provinces or counties) that expanded during the years prior to the U.S. Intervention, with rural zones still insufficiently served.

This lack of sources hinders an accurate assessment of the expansion set in motion by the Occupation government. However, it can be stated, without fear of misrepresentation, that the growth of the rural school was significant, perhaps not to the degree claimed by Calder, for whom "the largest growth occurred in rural areas . . . in 1916 there were, perhaps, 30 functioning rural schools . . . by 1920 there were 647 . . . ," an appraisal that overstates the case, as the first number had been easily surpassed by 1915 simply by the addition of rudimentary schools in the provinces of La Vega and Santiago.

A clear trend in hiring women teachers began once schooling became mandatory, and by the 1920s an unmistakable predominance of women in the national teaching corps was consolidated. In 1881, the overall participation of women in the teaching field was 32 percent; by comparison, by 1920 their relative numbers had jumped to 64 percent.

Although table 8.1 (based on regional payrolls for teachers) regrettably lacks information for counties such as Santo Domingo, the capital, whose greater school development contributed heavily to the national pattern of gender distribution of teaching positions, it permits us to observe interesting regional differences with respect to the feminization of teaching.

As can be seen, around 1920 there existed a definite trend toward the *feminization* of teaching in the North where, in many counties, women's participation exceeded 65 percent, and in the case of Samaná, reached more than 90 percent. Meanwhile, in the rest of the country, a contrary pattern of male predominance existed, except in certain isolated cases such as Barahona and Cabral. In this way, in

Table 8.1 Participation of men and women in teaching positions by county* 1917–1923

County	Women	Men	% Women
North region			
Santiago*	109	52	68
La Vega	17	8	68
Jarabacoa	3	8	27
Moca	15	8	65
Puerto Plata*	33	17	66
Samaná	10	1	91
Northeast region			
Montecristi	5	6	45
Dajabón	—	8	—
El Cercado	1	10	9
Las Matas de Farfán	2	7	22
Restauración	1	4	20
Eastern region			
La Romana	12	11	52
San Pedro de Macorís	8	13	38
Southeast region			
Barahona	14	3	82
Bánica	—	10	—
Comendador	1	9	10
Cabral	7	2	78
Duverge	3	7	30
Enriquillo	8	2	80
Neyba	2	10	17
South region			
Azua	14	23	38
San Juan Maguana	4	16	20
San José de Ocoa	2	7	22

*Provinces. The province of Santiago includes the counties of Esperanza, Jánico, Peña, San José de las Matas and Santiago; Puerto Plata includes the counties of Bajabonico, Altamira, Blanco and Puerto Plata.

Source: Based on Intendants' Reports to the National Service of Public Instruction.

contrast to the county of Samaná, which became almost synonymous with feminization, other counties existed such as Dajabón and Bánica that were totally masculinized, as well as a large number of counties where rates of women teachers were lower, including those where the overall rate in 1881 of 32 percent still prevailed, and others such as La Romana, Montecristi and San Pedro de Macorís where the rate was slightly higher than in 1881.

Without a doubt, it was the counties of the so-called Valle del Cibao region (in particular, those in the provinces of Santiago, La Vega, Moca, and Puerto Plata) that display a defined and significant tendency toward feminization. At the turn of the nineteenth century, El Cibao, doubtlessly as a consequence of the development of the tobacco market,[30] had the highest number of schools in the country. After 1880, the increased opening of schools in the Valle was in response to the need of local government and individuals to acquire literacy and math skills, or grew out of initiatives by workers to educate themselves.

The Feminization of Teaching in the Valle del Cibao

How and why did Valle del Cibao schools become feminized? With the exception of Puerto Plata, which in 1881 exceeded the general rate for women's participation of 32 percent, the remaining counties were below the national average at that time.

Part of the feminization process is due to the expansion of education to girls and women workers in urban contexts. Toward 1917, the schools in the provincial capitals were significantly transformed. High school education, for example, had become diversified with the opening of Normal Schools. While in 1881 only Santiago had a Normal School, in 1917 the cities of Moca, La Vega, and Puerto Plata had joined the Normalist movement, and had their own Normal School for women and men. The *Escuelas Graduadas Completas* and the *Escuelas Graduadas Modelo*, the first a successor to the *Colegios Centrales*, and the second designed to provide instruction for trades such as bookkeeping, etc. had also been inaugurated in these four cities, offering separate courses for men and women.

Primary schools also grew significantly during the years 1881–1917. Education for girls brought about an increase in women's teaching positions, due not only to the increased number of girls' schools but also to a tendency for mixed schools to hire women as teachers.

However, when examining the increase in female teaching positions, even if we consider the additional spaces that were implied by the expansion of primary education in the county seats (which by 1917 would have a primary school for boys and another one for girls, thus doubling the number), the main component of the feminization of teaching in the Valle del Cibao was the development of rural education, an area in which women had scarcely participated in the nineteenth century.

Once again utilizing information referring solely to the Provinces of Santiago and Puerto Plata to illustrate the overall situation of the Valle del Cibao, I will simply note that compared to the approximately

15 rural schools and the three, four or five women teachers working in these in 1881, by 1917, the year the Law of Mandatory Instruction took effect,[31] rural women teachers were already relatively widespread in the countryside, as seven of ten teachers working in the 116 rudimentary schools that year were female.

What caused these high rates of feminization? A tentative response must necessarily distinguish between the separate realms of the rural and the urban in attempting to explain why those women sought to be employed as teachers and why the system availed itself of their services.

FEMINIZATION IN THE URBAN CONTEXT

In the expanding job market for women teachers in the urban areas of the Valle del Cibao, one can see a growing influence of a middle class linked to the export of agricultural goods, and a parallel demand for a cultivated and broad education for its daughters who would one day marry or further their studies outside the country. Educated teachers were needed for educated girls. But at the same time, the expanded market for teachers may have responded to the increased number of working-class girls who were hired to work in domestic or business settings, or who were forced off the streets into schools by laws that made education mandatory. Teachers, some of whom were only partly educated themselves, were required to teach morals and instruct working-class girls in sewing, domestic economy, and the *habitus* for work.

An analysis of names and family ties on teacher payrolls for 1917 allows us to make certain inferences and speculate about the motives and social conditions in the *disposition* of those women who joined the teaching field. In the city of Puerto Plata, for example, more than a third of the surnames on a list of 21 women teachers are of foreign origin: Finke, Schild, Miller, Ashton, Lithgow, Roger, Pierret, Archambault. On similar lists for the city of Santiago de los Caballeros, the foreign surnames are not as prevalent (one finds Lithgow, Monony, Chamberlain, Smester, Wynns, Knipping, Giralt, Plá, Mascaró, etc.), but a significant number of surnames appear which reveal the presence of the educated and powerful *criollo* elite of the city, among them, Hereaux, García Godoy, Reynoso, Bidó, Hungría, Franco, and Espaillat.

The immigration of foreign families was a consequence of the broad contact with the world through the tobacco market. Lithgow, for example, was a surname linked to the development of the network of roads essential to the transportation of goods. It was also associated with the political power in the region, as one of its members, General Federico Lithgow, was governor of the province of Puerto Plata during Hereaux's administration. Teachers for the elite emerged from these families and reflected their cultural status through the teaching positions they held: Mary Lithgow was the principal of the *Escuela de Párvulos* in Puerto Plata, the first kindergarten in the country to use Froebel's methods, and her brother Juan was the principal of the *Escuela Modelo de Varones No. 5*, a boy's school in Santiago.

Isabel Finke, Carmela Archambault, Delfina de Schild, Mercedes Schild, Alicia Pierret, and María Rogers were principals in six of the 15 urban primary schools that existed in Puerto Plata in 1917, while Eleonora Cray and Lucinda Smith taught specific sections of dress-making and needlework classes. Daughters and wives of the second, or at times third, generation of European migrants to the country were women who no doubt possessed rare knowledge in areas such as foreign languages, which were useful for overseas trading. They dedicated themselves to educating the children of the elite who would later pursue their studies in foreign countries.[32]

These teachers, whose social ties are easy to trace because of the uniqueness of their surnames, show a certain family specialization in education. Together with the Schilds, mother and daughter, and the Lithgow family, this family constitution is also clear in the case of the heirs to the regional *criollo* elite, some of whose surnames denote their origins, such as Espaillat, Reynoso, Hungría, and Hereaux.

The Hereaux sisters taught in elementary schools for girls in Santiago. The Reynoso family, who were possibly related to Manuel de Jesús Peña y Reynoso, the founder of the "Sociedad Amantes de la Luz" (the cultural association in Santiago de los Caballeros), included several sisters who worked as teachers in that city in and around 1917. Members of the Espaillat family, clearly related to Francisco Ulises Espaillat, the most distinguished representative of enlightened rationalism in the country, were teachers in Santiago, La Vega, and Moca around 1920.[33]

What motivated these children of the elite to seek out poorly remunerated employment as teachers? It is difficult to determine an answer. Teaching could offer them a prestigious status, for tutoring the children of the elite made such a status possible. However, it is not unlikely that they also had a strong philanthropic or idealistic bent. This may well have prompted the descendents of Espaillat to follow in the footsteps of a man whose idealism led him to establish, along with Peña y Reynoso, the *Liga de la Paz*, a political, patriotic, and philanthropic society.

A certain cultural *disposition* allowed these urban women from Cibao to embrace teaching. Whether from the middle or upper class, the educational system welcomed these women for the express purpose of educating girls or teaching in mixed schools.

FEMINIZATION IN RURAL CONTEXTS

Other factors seem to have been at work in the case of the rural feminization of teaching. In the countryside, the feminized pattern acquired more accentuated traits. By 1920, the overall average rate of female participation in teaching jobs exceeded 64 percent in parts of certain counties within the Valle del Cibao. In the rural areas of the province of Santiago, the feminization of teaching exceeded the overall average rate, even reaching 100 percent in the county of Peña, where all the rural schools were run by women teachers. Moca, Puerto Plata, and Salcedo were counties where rates of feminization were also above the average 64 percent in the Cibao region. At the same time, instances of masculine predominance in teacher hiring can be observed within this same region, notably in Altamira, Bajabonico, Bonao, and Jarabacoa, counties that, despite their proximity to feminized zones, approached the masculinized pattern of teacher employment that prevailed in the rest of the country.

This salient factor in the situation of certain rural parts of the Cibao region (which quickly, and one could say, almost immediately assumed a feminized pattern) once again poses questions about the specific conditions for hiring teachers in the region, and about the elements underlying the *social availability* of those women who were hired as rural teachers.

As was shown, at the time that primary school education became mandatory, the educational system faced an enormous gap between the number of trained teachers and the demand generated by the new law. This gap increased because few teachers were interested in leaving cities and towns to live in isolation in the surrounding countryside for salaries which, in the time-honored tradition of teaching, were at the lowest rung of the salary scale.

Like other of his colleagues, around 1920 José Dubeau, General Inspector of the Province of Santiago, described in the following terms the circumstances that were prevalent in his district with respect to the hiring of teachers for rural schools:

Schools in the countryside leave much to be desired, not only because of the meager pay given to their teachers but also for the debatable fitness of these same individuals, in most cases. . . . Given the present state of affairs, this evil has no remedy, because it is not easy to secure competent teachers willing to leave the place where they live to go inland, where they lead a life that is not in keeping with their likes, traditions, and natural tendencies. . . . given the pittance they earn, it is almost necessary to entrust this mission to the first or, one might say, to the only one who applies.[34]

Dubeau's description does not exaggerate the great number of rural schools that almost certainly had as their teacher "the first or, one might say, the only one who applie[d]." Despite the fact that in several rural areas women teachers recovered an existing educational tradition that had begun at the end of the nineteenth century, the recruitment of Dominican teachers for rural schools is characterized in these years by its pronounced instability. The low salaries and the frequent interruptions in pay, the political clientelism that affected the hiring of teachers, the constant relocations while in service, and other mechanisms of school financing and everyday factors made teacher desertion a factor in the frequent stoppage of school operations.

Some evidence can be gleaned about the traits of teacher mobility at that time to give credence to Dubeau's opinion. In this regard, it is interesting to analyze such factors as the continuity of service of women teachers in the years 1918–1930. The analysis of the number of women who remained in their teaching positions over time reveals an important process of mobility in the rural schools of the Cibao region. It is possible to say, for example, that only one of ten rural women teachers who worked in Puerto Plata in 1918 continued in that position 12 years later; that in the course of a decade four of five women teachers working in Moca in 1920 had left teaching; and that only one of five women teachers who served in rural schools in La Vega in 1925 remained in their positions five years later. This pattern of mobility among the educational system's teaching staff seems unstable when compared to the life circumstances that affected the trajectories of teachers in subsequent years, and was the consequence of the factors mentioned earlier, such as changing political clientelism,[35] and re-locations while in service.

One could argue that the mobility of those who disappeared between the first payroll and the later one implies their almost definite exclusion from the system, since they did not appear to have incorporated themselves in other administrative districts within the National Service of Public Instruction. That is, there are no signs of any significant mobility between regions. Only a small number of teachers circulated from one school to another within the rural district where they had joined the service, and the cases in which teachers move between counties, school districts or provinces are few and far between.

Besides its transient nature, the insertion of many of the teachers during those years had uniquely *local* features. Inspector Dubeau's description is certainly validated by evidence showing that jobs were awarded to whomever was willing to teach, even for a short term. However, before probing more deeply into the sociology of teacher hiring on the local level and its tendency toward feminization, it might be useful to point out some aspects of the tradition I referred to earlier of women who taught in the Valle del Cibao countryside in the last decade of the nineteenth century.

Regarding the careers of these women teachers, some information is provided by a diagnostic test given in 1930 for the purpose of weeding out teachers to make room for a generation of graduates from the Normal Schools and those with a degree in primary and secondary school teaching, who had applied pressure to obtain teaching positions. A resolution appears to have been achieved through a very early political alliance between the teachers and General Rafael L. Trujillo.[36]

The results of this diagnostic test describe teachers such as Blasina Gómez Betemit, who began her teaching career in approximately 1903, and for more than 27 years headed the primary school in Sabana Grande, in the county of Peña, which was not far from where Mercedes Vda. Torres (widow of Torres) had been teaching since 1904. It also lists teachers such as Teodora Vda. Knipping, who began teaching well before the Military Occupation in Gurabo, a beautiful area close to the city of Santiago, where on one country estate— perhaps hers—Don Guillermo Knipping's famous German grand piano was played. Or teachers such as María Teresa Calderón, who had taught in Junta de los Caminos since 1912, when the school was run by the municipality, and who survived as a teacher despite the ups and downs of the public treasury (depending on the year, she was either a municipal, national, or private teacher). She also provided space in her house for classes given by Señora Vda. de Álvarez, a teacher in the county of Santiago since 1897.

The so-called "viudas maestras" or "widowed teachers" continued to teach until the days of the Trujillo teacher replacement. These teachers were typically commended by their Inspectors more for their pleasant, patient, loving, and maternal manner than for their devotion to the new objective education.

This image of a rural educational system, one that was heavily dependent on the sociological context in which it operated and inclined to hire its teachers locally, is corroborated by other facts. It was not only the lack, absolute or relative, of training and the low salaries that brought about the hiring of "the first or, one might say, the only one who applie[d]," but also the absence of schools and the resources to build them that determined, in great part, the profile of the teacher with whom the policy to increase primary school education in rural areas was launched.

In 1918, José Francisco Camarena, Inspector of Public Instruction in the County of Bajabonico, reported the following to his superiors:

> From notes kept during inspections, as well as from the information about the locations where the public schools have been established . . . it can be observed that all these sites are owned, if not by the school principals, then by someone known to them, which creates the following dilemma: either the schools in which the principal is being replaced (or should be replaced) be temporarily closed, or adequate sites be constructed as soon as possible so that the change can be implemented.[37]

As Camarena points out in another report, this dependency on sites that were provided by the school principals or by people known to them meant that "the organization that wishes to introduce Public Education . . . will run into serious problems that interfere with the proposed objectives."[38] Replacing a principal, male or female, for the purpose of giving the job to someone more pedagogically suited could even jeopardize the continuity of the institution on the local level.

Examples can be cited of women who provided spaces for schools, whether it was because they owned the space or it had been ceded to them by a landowner with whom they had a relationship, as in the above-mentioned case of María Calderón. In La Judea, in the county of Montecristi, an incident revealed the nature of certain local relationships that paved the way for women's participation. In 1930 the School Inspector wrote:

> Regarding the permanent closure of the rudimentary school in "La Judea" and the re-opening of the school that starts operating once again today in Copey, this office advises . . . that the procedure was specifically intended to open a section in Copey, where it belongs, as it is the most important town in this county. With respect to the petition of Sr. Israel Álvarez to the Honorable President of the Republic and to

the General Superintendent of Education . . . for the purpose of maintaining the school in La Judea, this inspection affirms without fear of equivocation, that Sr. Álvarez is pursuing his own personal gain . . . because the former principal of this school is a family member, and moreover, is the one in charge of overseeing the laborers that this same individual has in La Judea, whose lands are almost entirely in his ownership.[39]

According to other reports, the locale for the school in La Judea had been rented by Sr. Álvarez starting in 1897, and Señora Carmen García had served as its teacher ever since. The image of a rich farmer turned benefactor who reduced rents for the county, provided support for free education and, perhaps, ordered the teacher to oversee his interests, speaks to us of a kind of female participation that was mediated by the local elite. Local education stood to benefit from the relations between the teacher and the local political party.

Along with the scarcity of candidates who were suitable for teaching, and the lack of schools, a third factor that limited the possibilities of the educational system to determine who would be hired as a teacher (and control, as a consequence, the conditions under which the curriculum was transmitted) was a by-product of local and regional politics. During the early years of the expansion of obligatory schooling, the wishes of provincial inspectors and supervisors frequently collided with local interests over the choice of the individual to be hired as teacher in the schools.

In essence, the struggle to determine *who* had the authority to recruit teachers (was it the municipal or provincial authority or the Inspector?) revealed not only problems of a *jurisdictional* order, commonly found in many nations with the introduction of a centralized curriculum, but also exposed the clientelism of the Dominican state apparatus. Although the local authorities had been stripped of their jurisdiction over school administration by the 1914 Code, and correctives were introduced to the educational patronage system by the 1918 Organic Law of Public Education,[40] the resistance of local and regional powers significantly limited the power of the regional Supervisors of the National Service of Public Instruction to give teaching jobs to the individuals who, in their opinion, were best suited.[41]

In conclusion, then, with respect to the hiring of teachers, the three above-described sociological factors weighed heavily: the *disposition* to accept a teacher's unreliable salary, the local control of spaces, and the local political control over the distribution of employment. The choice between curriculum and society in the shaping of the educational system was clearly resolved, at least during this early formative phase, in favor of the latter.

AN INTERPRETIVE HYPOTHESIS REGARDING THE FEMINIZATION OF TEACHING IN THE CONTEXT OF RURAL CIBAO

The three sociological factors described above were, however, so to speak, gender-neutral, because they can be observed in the employment of both men and women. The question I posed earlier about the reasons why feminization occurred in certain counties in rural Cibao is still pending, as all these situations (ties to local power, control of space, *disposition*) could also be experienced by men. In what follows I will attempt an answer.

The counties with high female participation in primary schools in 1917 were the most developed in Cibao. From the early nineteenth century, the district of Santiago was the central hub for regional connections between the campesinos and the international market in tobacco, coffee, and cacao, which later expanded in two directions: toward the East (where the fertile lands of Moca and Salcedo were located) and in the North, toward Puerto Plata, through whose port a significant portion of commercial imports and exports moved.

In essence, the rural Cibao society was composed of campesinos with small landholdings. It was through their work that the wealth was created that sustained an extended network of merchants and middlemen and brought about the development of the entire region. Initially specializing in the tobacco trade with Germany, by the turn of the nineteenth century the campesinos of Cibao intensified their commercial ties through the additional production of coffee and cacao, crops whose market demanded a continuous expansion of the agricultural frontier. While this was not a favorable economy with full employment, what is certain is that the campesinos' small output must have made them intensify their processes of work as a survival mechanism within an economic order where they were subordinate to mercantile and usury capital.

In his book *Los Campesinos del Cibao*, Pedro San Miguel analyzes in great depth the dynamic of rural development in the region, describing the means by which campesinos saw themselves increasingly subjugated to mercantile trade.[42] Two aspects of his historiographic analysis help strengthen the argument that the observed feminization of teaching was connected to the significant transformations resulting from the intensification of trade rather than from the popular preference for women or from pressure by the curricular designers to feminize teaching.

San Miguel demonstrates the very demographic structure of Cibao, exhibits a low rate of maleness. In other words, there were more women than men, a pattern that was especially pronounced in the cities (in Santiago de los Caballeros, for example, the gender ratio was 86 females for every male in 1916). To a lesser degree, women also outnumbered men in rural zones. As a demographic feature, the imbalance in the male/female ratio could have framed the general context of the job market for teachers.

However, one factor of greater importance in the feminization can be inferred from the transformations that affected the life of *men*. In this respect, the general dynamic of the period was not only toward a more intensive farming of small plots of land. Men, both young and old, also began to migrate from one rural zone to others, whether to cultivate new lands for their own crops or to join the regional market of wage-earning semi-proletariats that had already begun to form.

The market dynamics also tended toward the relentless exploitation of male work. Laws were passed that burdened workers, such as the *Ley de Caminos*, a law that forced laborers to work on jobs related to the spread of the communication networks that were required by the expansion of the market. In San Miguel's analysis, as a consequence of this law, the campesinos of Cibao had to intensify their own processes of work, whether by working directly on the construction and maintenance of roadways or by paying a fee that exempted them from direct involvement in the process.

Obliged to work more in order to survive in this structure of commercial exaction, and forced to contribute to the creation of conditions for the expansion of the market (roads, irrigation canals, etc.), the campesinos of the Cibao region had to redouble their productive energies during this period to endure as a class. This increased demand tended to fall on men's shoulders.[43]

Characteristically, it is the counties that had the greatest trade exploitation and the most construction of transportation infrastructure that have the highest rates of feminization in the Cibao region. If the feminizing trend was derived from ideographic and representational aspects alone, it should have been equally evident in counties where agriculture was not as integrated into the market, but this is not the case—in these rural schools the traditional male dominance is found.

In line with this reasoning, an explanatory hypothesis of the process of feminization in Cibao must examine the social conditions that, in an indirect way, blocked men's access to the teaching market, at least in its rural sector where, as has been seen, conditions existed that made teacher recruitment a characteristically local process.

THE FEMINIZATION OF TEACHING IN THE DOMINICAN REPUBLIC

Given that seven out of ten teachers in the public education system were women by the mid-1930s, the Dominican teaching field had by that time acquired a clear pattern of gender distribution that turned national education into an essentially feminized system. Policies of the 1920s regarding mandatory education generated this pattern, which has continued to characterize the Dominican national school until the present day.

As I have argued throughout this chapter, up until that decade, the institutional actions of the State with respect to the curriculum had little to do with the feminization of Dominican teaching. At the time the nation consolidated its educational practices, its main impulses derived from processes associated with international migrations, demographic changes and the social and cultural history the relatively young republic had inherited.

The cultural history of representations and gender practices in Dominican schools no doubt owes much to the weaker foundations of the state, unlike the case in other nations, where the control of curricular transmission was much stronger. A central factor in Dominican history—and a subject requiring further study—are the processes that made the Caribbean a site of intense international migrations. A wide spectrum of traditions and practices came together in the Dominican national space regarding the participation of women in education. Into this context of international migrations can be added a social space weakly regulated by public power, making the Dominican experience a unique site of convergence, where the Anglo-Saxon cultural universe blended with the Spanish colonial matrix along with possible variants introduced by the British, the Dutch, the American, and the French.

In order to understand how women participated in Dominican teaching and the origin of the process of feminization, it is necessary to consider how the cultural history of the Dominican Republic was shaped by the international migrations that characterized the Caribbean during the nineteenth century.

As I have argued, the conditions of the feminization process were substantially modified from the moment the State set out to control the curriculum and make education mandatory. The implementation by the Occupation Government of a policy to educate the masses gave rise to an important process through which women were inserted into Dominican schools. The expansion of employment for female teachers was notable, above all in those areas of the country that were undergoing urbanization and productive changes, such as in the Valle del Cibao, a region that not only made a disproportionately high contribution to the feminization of the national teaching force, but was also one of the first areas of its historic expression.

In examining how the process of feminization developed in the Valle de Cibao, it is necessary to distinguish the dynamics of an urban setting from those that operated in the rural context. In the cities of the Valle de Cibao, the opening of teaching spaces for women corresponded to the expansion of the middle class and the growth of services. These urban women teachers, many of whom were from the wealthy sectors of their societies, found conditions that were suitable for their insertion into the labor market because of the growth of girls' education and mixed schools. However, this logic did not seem to apply to rural women teachers. For rural areas, the analysis of women's insertion into this market must also incorporate factors that blocked men's access to the teaching field.

NOTES

1. In 1881, the overall participation of women in the teaching field was 32 percent; by comparison, by 1920 their relative numbers had jumped to 64 percent.
2. With respect to this tension, there are several recent studies on the history of education that focus on the mechanisms through which society imposed its own modalities on the State's project of institutionalizing schools. See, for example, Mary Kay Vaughan, *La política cultural en la revolución. Maestros, campesinos y escuelas en México. 1930–1940*, (México: SEP, 2000); María Bertely Busquets, *Historia social de la escolarización y uso del castellano escrito en un pueblo zapoteco migrante* (Ph.D. Dissertation, Universidad de Aguascalientes, Mexico, 1998).
3. Such as Ulises Francisco Espaillat, Pedro. F. Bonó, Fernando A. Meriño, and Gregorio Luperón, among others.
4. Código de Educación Común
5. It was surely the unstable institutional infancy of the Dominican state that explains the apparent absence of a discursive production on the subject of gender in education and the absence of specific designs such as those documented by Sonsoles San Román, *Las primeras maestras. Los orígenes del proceso de feminización docente en España* (Barcelona: Ariel, 1998) in Spain.

6. The rudimentary schools (*las escuelas rudimentarias*) were established in rural areas and were so named because they imparted classes in the "rudiments" of learning: math, reading, and writing. Their curriculum was a synthesis of the elementary school curriculum used in the cities.

7. The Organic Law of Public Education [*Ley Orgánica de Enseñanza Pública*] of 1918, decreed by the U.S. Military Government, weakened the former division by gender in primary schools that was established in 1914 by the Public Education Code. The Organic Law did not explicitly refer to mixed education; it was implied, however, because differentiation by sex was not re-established. See Gobierno Dominicano, *Colección de Órdenes Ejecutivas. Años 1917–1922* (Santo Domingo: Imprenta de Vda. García, 1968), pp. 354–379.

8. Servicio Nacional de Instrucción Pública

9. The only published historical study I know of is by Ramón Morrison, *Historia de la educación en la República Dominicana (Desde sus más remotos orígenes hasta 1900)* (Santo Domingo: Editora Taller, 1995).

10. Ley de Instrucción Pública.

11. A *común* is the political and administrative division equivalent to a county into which the Dominican Republic is divided.

12. Morrison, *Historia de la educación en la República Dominicana*, pp. 55–57.

13. Twenty years later, the educational situation of the country had hardly changed, as can be gathered by the summary presented by Bonó. See the report he prepared as Minister of Justice and Public Instruction in 1867 in Emilio Rodríguez Demorizi, *Papeles de Pedro F. Bonó* (Santo Domingo: Academia Dominicana de la Historia, Editora del Caribe, 1964), pp. 144–150.

14. The list is reproduced in Morrison, *Historia de la educación en la República Dominicana*, p. 71.

15. See Harry Hoetink, *El pueblo dominicano 1850–1900. Apuntes para su sociología histórica* (Santo Domingo: Ediciones Librería La Trinitaria, 1997), p. 285.

16. Hoetink's study is pioneering with respect to the subject of the "English schools." However, as I mention, the school system for freed slaves and their descendents has not been sufficiently studied. It can be shown that by 1900 there existed approximately seven schools of this type in Samaná. As the years passed, the system would grow. Hoetink, *El pueblo dominicano 1850–1900*, pp. 50–53.

17. I use the term "Yankee" to indicate that these were freed U.S. slaves who came from northern cities of the United States such as Philadelphia.

18. The census is reproduced in Morrison, *Historia de la educación en la República Dominicana*, pp. 205–221.

19. Neighbors of Estancia del Yaque to Horacio Vásquez. Estancia del Yaque, September 15, 1928. Archivo General de la Nación, Provincia de Santiago, 1929, folder no. 35, unclassified file.

20. Cited in Hoetink, *El pueblo dominicano 1850–1900*, pp. 236–237.

21. Michael Apple has analyzed the class connotations implied by the distinction between "ornamental sewing" and "practical sewing" within the British school system in the nineteenth century. See Michael Apple, *Maestros y textos. Una economía política de las relaciones de clase y de sexo en educación* (Barcelona: Paídos, 1989), p. 77.

22. It was a custom for widows to append "Viuda de" followed by her deceased husband's surname.

23. In Latin America a woman typically keeps her maiden name and adds to it her husband's surname preceded by "de."

24. Morrison, *Historia de la educación en la República Dominicana*, p. 130.

25. Ibid., p. 275.

26. Hoetink, *El pueblo dominicano 1850–1900*, pp. 52, 241–242.

27. See Mario Concepción, "Geografía del apellido dominicano," *Revista Eme-Eme*, Santiago: Universidad Católica Madre y Maestra, VI (March–April 1978): 28–47.

28. E. Zúñiga, Educational Inspector for the school district of Samaná and Intendant of Education in the Departamento Norte. Samaná, December 9, 1919. Archivo General de la Nación, Inspección Samaná, 1919, file 1/735.

29. See Bruce J. Calder, *El impacto de la Intervención. La República Dominicana durante la Ocupación Norteamericana de 1916–1924* (Santo Domingo: Fundación Cultural Dominicana, 1989), p. 51.

30. The Cibao region, the birthplace of Espaillat, Bonó, and Luperón, was the cradle of the nation's enlightenment with its greater social development and its cities, which were as old and as sophisticated as the capital, Santo Domingo.

31. This law *La Ley de Instrucción Obligatoria* made education between ages 7 to 14 mandatory, and was put into effect by the Executive Order No. 114 issued on December 29, 1917 by the U.S. Military Occupation Government. The Law of 1917 was thus the first law intended to achieve mandatory education.

32. I find it of interest to recount the anecdote told by Hoetink about the arrival in Santiago of a German grand piano that had been ordered by Don Guillermo Knipping at the end of the nineteenth century. The legendary piano was carried on men's shoulders from Puerto Plata to Santiago. This epic feat—so reminiscent of the film *Fitzcarraldo* by Fassbinder—speaks of the cultural signifiers present in Santiago society. The woman who was most probably his widow—the teacher Teodora Vda. Knipping—taught in that county in the early years of the twentieth century. Like Knipping's widow, other women of foreign origin served in rural primary schools.

33. Julio G. Campillo Pérez, *Francisco Espaillat y el desarrollo del Cibao* (Santo Domingo: Instituto Dominicano de Genealogía, 1985).

34. José Dubeau, Provincial Intendant of Education. *Informe*. Santiago, December, 1918. Archivo General de la Nación, Education Branch. Province of Santiago. File 1/682.

35. During the years 1918–1930, excluding the interlude of the U.S. Occupation and Military Government, clientelism was a significant factor in teacher hirings.

36. One of the first actions the Trujillo regime took was to order inspectors and intendants of the National System of Public Instruction to evaluate the pedagogical merits of rural teachers. This led to a series of dismissals of teaching personnel. It is interesting to detect the way in which a new *trujillista* political clientele, made up by a strata of individuals who held teacher training credentials, tended to displace many long-standing teachers and others who, in the comings and goings of political alliances, had been placed at the head of primary schools.

37. José Francisco Camarena, Inspector of Public Instruction for School District 31 to the Intendant of Education in the Departamento Norte. Bajabonico, December 7, 1918.

38. Ibid., Altamira, October 5, 1918.

39. District Inspector to the Intendant of the Departamento Norte. Montecristi, February 6, 1931. Archivo General de la Nación. Province of Montecristi. Folder 11, Inspection.

40. Articles 10 and 11 of this Law are intended to limit political-propagandistic uses of teaching. See *Colección de Órdenes Ejecutivas. Años 1917–1922*, 355–357.

41. José Dubeau, Education Intendant General to the Superintendent General of Public Instruction. Santiago, March 3, 1916. Archivo General de la Nación. Education Division. File 1/663.

42. Pedro L. San Miguel, *Los campesinos del Cibao. Economía de mercado y transformación agraria en la República Dominicana 1880–1960* (San Juan: Editorial de la Universidad de Puerto Rico, 1997).

43. See San Miguel, *Los campesinos del Cibao*, pp. 257–300. Orlando Inoa has also analyzed the impact of enforced labor on the campesino economy of the Valle del Cibao. Orlando Inoa, *Estado y campesinos al inicio de la Era de Trujillo* (Santo Domingo: Ediciones Librería La Trinitaria, 1994).

CHAPTER 9

Teachers of Yesteryear: A Study of Women Educators during Porfiriato

Luz Elena Galván Lafarga

INTRODUCTION

The search for images of women teachers of yesteryear is the principal aim of this article. The letters sent by of these teachers to President Porfirio Díaz (1876–1910)[1] open a door to the past. We are interested in looking through that door at the women who wanted to be teachers, or perhaps already were, as they appropriated and adapted the rules that emanated from the powerful head of state. This study will be about the women to whom orders were given, about women "without power," and about relationships that were not merely from the top down, but which also included a "complicated game of checks and balances" that "allowed a specific regime to endure."[2]

This study draws us closer to the everyday cultural practices of several women teachers of yesteryear, to their training, to their activities as teachers, to their successes and failures, to their desires for self-improvement, to their salaries, to their daily needs, and to their body, mainly by studying their illnesses and hygienic practices.

During the Porfiriato,[3] women teachers appropriated and adapted the dominant culture, but at the same time, they "re-elaborated" it, giving rise to a culture that was qualitatively different. It is thus a question of two cultures: the dominant and the subordinate, each influencing the other, above all through the opinions of different women teachers who, at the same time, negotiated the spaces of interaction and arrived at a "harmonic coexistence."[4] This led several teachers to develop strategies that allowed them to introduce new ideas into the Porfirian educational system.

In addition, this study will investigate how these women teachers played the role of "cultural intermediaries," understood to mean those individuals who leave the community and for a period of time live in other sociocultural contexts, where they learn "other habits, traditions, and ways of thinking" that they put into practice upon their return.[5]

These women teachers formed a part of the educational project implemented during the Porfiriato. However, it must be noted that this project was the result of a long period during which economic, political, and social reforms were attempted, dating back to the birth of Mexico as an independent nation in 1821. Within the area of social reforms, the educational policies carried out during the nineteenth century are noteworthy.

EDUCATIONAL POLICIES

In 1823 a project of public education was launched, which reflected the desire of the governing class for education to reach a public of both girls and boys, as well as adolescents and adults. However, the political situation faced by our country did not permit this ambitious plan to be realized.

Nevertheless, some states were concerned about teacher training. During the nineteenth century up until 1921, when the Ministry of Public Education was created, each state managed its own educational system independently, in keeping with its budget.[6] States such as Zacatecas, Chihuahua, Tamaulipas, Nuevo León, Veracruz, Jalisco, Michoacán, and Oaxaca united their efforts with those of the Compañia Lancasteriana, in order to establish schools where future women and men *preceptores*.[7] would be trained.

The Compañia Lancasteriana was named after one of its founders, the Englishman Joseph Lancaster, who established the company with Andrew Bell. The Lancaster system was introduced to Mexico by Manuel Codorniú, who arrived in the country with the viceroy O'Donojóu in 1821, and started the newspaper *El Sol*. The Compañia Lancasteriana, which was financed by private capital, promoted free primary school education for the poor classes of Mexico. Its first school, "El Sol," opened its doors in 1822; the second was called "Filantropía." The teachers who passed its courses were certified to teach using the monitorial system,[8] in which the teacher did not directly give classes to each of the hundred or more students who attended, but instead depended on the more advanced students, who were called monitors, instructors, or decurions. These children had previously been given instruction by the teacher in how to lead small groups of ten students. They taught reading, writing, arithmetic, and Christian doctrine. The Company established branches in provincial capitals, and although it did not succeed in opening the number of schools it desired, it contributed significantly to the earliest efforts to educate the nation. José María Luis Mora commented that, in the first years following independence, primary education in Mexico had not been completely established, but it had expanded in an "astonishing" way throughout the entire Republic. The Lancaster Company continued to exist until 1890, when it was banned. By that time the Federal government had increased its support of municipal schools, and the Lancasterian teaching methodology was considered outdated.

Political problems faced by the new Republic, together with invasions by the United States (1847) and France (1862–1867) made it necessary for private institutions to offer public education in Mexico. Besides the Compañia Lancasteriana, the Sociedad de Beneficencia para la Educación y Amparo de la Niñez Desvàlida, the Asociación Artística Industrial and the Sociedad Católica Mexicana, among others, should be mentioned.

It was during the period of the Restored Republic (1867–1876) that the educational project of the Mexican State began. In 1867, the liberals defeated the empire of Maximilian, who had been installed by the French troops sent by Napoleon III. The educational policy developed by Porfirio Díaz during his long government thus had its precedent in a policy that dated back to 1867. That era was distinguished by the arrival of Benito Juárez (1867–1872) to the Presidency. From the outset of his administration, Juárez was concerned about the development of public education, and laws were passed aimed at increasing the number of free schools and educating women, among other reforms.

In its ideological aspects, the speech given by Gabino Barreda[9] in Guanajuato on September 16, 1867, had great significance. His speech avowed the positivist doctrine that the liberals brought with them to power once the Republic had triumphed. He expressed the philosophy that would be associated with this entire epoch, one that would end in 1911 with the downfall of Díaz and the onset of the revolutionary movement. Starting with Barreda, the idea of using education to transform patterns of behavior and bring about an economic and political modernization became a permanent characteristic of the educational policies in Mexico.[10]

What Barreda proposed was liberty, order, and progress. Liberty as the medium, order as the base, and progress as the end. For him, it was of utmost importance to preserve the material order, so that the nation could "walk on the path of progress and civilization."[11] He believed that the chaos that existed in society was a result of the chaotic mind of the Mexican; order had to be imposed through education. He maintained that all Mexicans had to possess a common background of knowledge, and that it should be of an encyclopedic nature. The means to establish this common background was through the positivist method. To successfully implement Barreda's program, education had to begin in primary school and all Mexicans had to attend school to "become orderly." Primary school was thus proposed as obligatory. Moreover, Barreda intended that education should be for both sexes, and declared that "all the reasons that exist to justify education by the State for males should be equally applied to female instruction."[12]

The ideological emphasis of positivism was distinct from liberalism: what was most important was not liberty, but order. As Barreda said, "yes to liberty, but a liberty that follows upon order."[13] The idea was to instill peace and order with the aim of achieving the material progress of the country; the new education would be shaped with this end in mind. Positivism was thereby introduced into Mexico not only as a philosophy, but also as an educational system.

While Juárez's government was being consolidated, educational reforms began. On December 2, 1867, a law was passed to reorganize public education in accordance with the principal features of positivism. This law insured that education would be secular by banning religious education in official schools. Article 1 decreed that "in the Federal District there shall be the number of primary schools for boys and girls as required by the population."[14] Article 5 stipulated that primary education be "free for the poor and obligatory in accordance with the law." It was believed that by making education mandatory, "the State would defend the rights of children against the selfishness and ignorance of parents." Also, the creation of an "educated teaching staff" to provide primary education was contemplated. There would be three categories of primary school teachers; all had to "pass the necessary examinations, and their category would depend on the number of subject areas that they passed."[15] One of the institutions that was inaugurated at that time was the Escuela Nacional Preparatoria, directed by Gabino Barreda himself.

Mandatory education in Mexico was a novelty; it was therefore imposed cautiously. It was declared obligatory beginning at age five, and rather than punish the children who did not attend, those who went to school were rewarded in various ways. However, it was emphasized that in order to be employed, parents had to show that their children were being educated. It was noted that "by 1869, almost all the states of the Republic had adopted the system of free and obligatory primary education."[16]

At the same time, the government also supported the Compañía Lancasteriana in its educational work. The Company still employed the monitorial system, even though the pedagogical literature had already begun to discuss the system of objective education.[17] To support the Compañía and to establish the new schools that were envisioned under the public education law, a public education fund was created into which capital derived from the nationalization of ecclesiastical properties was directed, as well as revenue from estate taxes. Moreover, the government obliged municipalities to create schools, and authorized them to invest part of its income to advance public education.

As far as school organization was concerned, the municipality was in charge. The existence of municipal schools dated back to 1821, when town and city governments were considered to be the representatives of the population, and it was thus their duty to oversee the needs of the public, education among them. In 1867 Juárez confirmed this idea when he ordered municipal governments to establish and provide funding for municipal schools. The municipality was thereby responsible for primary schools in the capital; these, however, were in very bad condition, described as "a deplorable situation of filth and neglect," and were overcrowded, since each school was run by a single man or woman teacher.[18] The inability of the municipalities to establish and operate an educational system made it necessary for them to rely on the Lancasterian schools, which continued to function.

However, despite Benito Juárez's efforts in favor of public education, the project could not be realized in its entirety because of the precarious state of the Public Treasury. As a result, it was not until the government of Porfirio Díaz (1876–1911) that changes began to be seen in the public educational system. By 1882, the municipal government in Mexico City had made notable progress: from 1845 to 1869 it had grown from three schools to ten schools, and then by 1882 jumped to 81 schools. It also established a system of civil service tests to hire new teachers and principals, in addition to establishing a Teachers' Academy and accepting public bids to write new school textbooks.

In that same year, a Congress on Pedagogical Hygiene was held during which questions were addressed relating to hygiene within school buildings, school furnishings, books, equipment, teaching methods, and distribution of work. Although this Congress did not take place at a national level, its ideas reached various states in the Republic, such as San Luis Potosí, where it was described in the Official Record.[19]

In 1888 an important law was passed that was intended to improve public education. Among other matters, it established that all official primary schools would be free, that no "persons belonging to any religion" would be accepted, and that the first four years of instruction would be "obligatory in the Federal District and the Territories,[20] for males and females between the ages of 6 to 12." It also decreed that the number of schools be multiplied: one for boys and one for girls for every four hundred inhabitants. Even though the municipalities had to support the schools, the federal government provided subsidies.[21]

Under the 1888 law, the State formalized the principles of a free, secular, and mandatory education, and at the same time increased its control over public education. This law is regarded as the "keystone" for all legislation dealing with primary education, as other laws were derived from it. Even though it covered only the Federal District and the Territories, over time it turned into a model that was copied by several states of the Republic. It divided primary education into elementary (four years) and superior (six years); it allowed children to be educated at home, but their studies had to be certified later by an official exam.[22] The federal government was in charge of formulating the program of study for elementary schools. The intellectual direction and educational materials for primary schools in the municipalities was left in the hands of the Ministry of Justice and Instruction, and to ensure that schools complied with the rules, various oversight bodies and cadre of inspectors were created.

The concept of mandatory schooling was reinforced by Justo Sierra[23] to combat the absenteeism among girls and boys in school. He commented that if a father didn't fulfill the obligation to send his children to school, then the State would take indirect measures to induce him to comply. A proof of writing skills was required by law to work as a State functionary, thus indirectly imposing a literacy requirement among the electorate.[24]

This law was quite ambitious, and reflected the longings of the first liberals who, like José María Luis Mora and Valentín Gómez Farías in 1833 and Benito Juárez in 1867, fought to achieve a free, obligatory, secular education. Unfortunately, these longings were realized only in theory and very seldom in practice. In fact, many boys and girls did not attend schools, either because they had to work to support their families economically, or because parents feared that their children would be infected with a contagious disease in the schools. In the nineteenth century, child mortality was very high due to the lack of hygienic conditions in public spaces, including the schools themselves. As a result, teachers complained constantly about high absentee rates among students.

With the aim of making education uniform throughout the entire country, Joaquín Baranda,[25] the Minister of Public Instruction, convened a Congress on public education in 1889. Under the leadership of Justo Sierra, sessions began at the end of that year, with representatives attending from each state. There were many subjects addressed at the Congress, including primary education, Normal Schools and secondary schools, programs of study, school buildings, adult education and rural education. All these subject areas could not be covered before the end of the First Congress, so a Second Congress was called for the end of 1890, which lasted until February 28, 1891. The most important subjects covered in these two Congresses focused on materials for primary education. The programs of study in nursery schools were formulated and teachers were left with the task of creating programs for primary schools; the program for upper grades had been outlined. Also discussed were the main characteristics of rural schools, traveling teachers, and schools for adults.[26]

Most importantly, the participants in the Congresses agreed that a uniform national system of "popular obligatory education" would be very advantageous, and an official school program was adopted as a result. After the Congresses concluded, several Mexican states passed laws which reproduced the federal school program.[27]

In accordance with a law passed on June 3, 1896, the primary schools that had formerly belonged to the Federal District and the Territories were nationalized. Thereafter, the Director General of Public Instruction was put in charge of 113 schools formerly under the direction of the municipal government of Mexico City. The schools, equipment, and furniture was organized and little by little the teaching staff was improved. One consequence of this nationalization was that both the number of schools and teachers were increased, in order to raise school attendance. However, despite these advances, the desired attendance figures were never realized, and the number of individuals who could read and write show a measurable increase, since many adults still did not go to school. There was much work to be done.[28]

In 1904, Porfirio Díaz announced that there were 523 primary schools (Grades one to six) in the Federal District and the Territories, with a population of 65,024 students, compared to 498 schools in 1903.[29] An important event during this period was the creation in 1905 of the Ministry of Public Instruction and Fine Arts, which replaced the Ministry of State and the Office of Justice and Public Instruction. The Ministry only had jurisdiction in the Federal District and the Territories. At its head was Justo Sierra, for whom the project of educating the nation was of utmost importance. He was aware that all the efforts that were made to progress would come to nothing unless the nation was first educated. However, the Minister of the Treasury, José Ives Limantour, did not share this view, and Sierra never obtained the budget necessary for the project of public instruction to succeed.[30]

In 1895, 14.39 percent of the population was literate; in 1900, it was 16.06 percent and in 1910, 19.74 percent.[31] When Díaz left the government, the rate of illiteracy was 80 percent. It was not until 1921 when a new Ministry of Public Education was created with an increased budget under the leadership of José Vasconcelos that the Mexican population was granted greater access to the educational system.

However, despite problems with school absenteeism and the lack of economic resources, the efforts to expand education during the Porfiriato did not end. Women and men teachers continued to be hired and sent to work throughout the entire country. Little by little, teaching became an important source of employment for Mexican women, many of whom had previously worked as teachers in Lancasterian or municipal schools in the different states of the Mexican Republic. For this reason, we ask the question: who were the women who collaborated on the project of public education during the presidency of Porfirio Díaz?

In Search of Images of Women Teachers during the Porfiriato

A few decades ago, Norwegian historian George Rudé[32] proposed that individuals must be recovered with their names and occupations along with their personal histories. This type of methodology corresponds to social history, which privileges the study of the common person instead of great men (which positivist history advocated). In this section the search for women teachers will begin. We will look at both the institutions where they received teacher training, as well as the public speeches given by statesmen of that era.

As was mentioned earlier, many kinds of women had entered the teaching field during the nineteenth century. In fact, before they were allowed to attend Normal Schools, there existed several institutions that prepared women to become teachers. Among these were the Secundaria de Niñas in Mexico City (which opened in 1869), the Instituto de Niñas in Durango (1870), the Liceo de Niñas in Aguascalientes (1878), the Liceo de Niñas in Guadalajara (1884), the Academia de Niñas in Morelia (1886) the Academia de Niñas in Nuevo León (1891), and the Escuela Normal de Artes y Oficios para Señoritas in the state of Mexico (1891).

Besides these institutions, Normal Schools were established in San Luis Potosí (1868), Guanajuato (1870), Puebla (1879), Nuevo León (1891), Tamaulipas (1890), Mexico City (1890), and Michoacán (1915). We also find that Normal Schools for both sexes were established in Nuevo León (1881), Querétaro (1885), Coahuila (1894), Jalisco (1904), Puebla (1906), Guerrero (1909), and Colima (1917).[33]

In some institutions, such as the one in Durango, girls enrolled at age 11 in a four-year course of study for a career as a *preceptora* of basic reading and writing. They would thereby enter the job market at age 15. We can affirm, then, that the incorporation of women into the teaching field was a consequence of "easy access," as Sonsoles San Román has described.[34]

These very young women would, on occasion, devote their entire lives to teaching. Some even spent four to five years as boarders at the Normal School, as in the case of scholarship students in San Luis Potosí.

This brief outline of the institutions and Normal Schools that trained women teachers leads us to the next question: why did women enter the teaching field? Part of the answer can be found in the public speeches of the era. In fact, during the Restored Republic, José Díaz Covarrubias stated that women were "the sort to educate children," as he viewed them as "gracious, sweet and pure," qualities that he believed made them more suitable than men for teaching.[35]

Covarrubias explained that an investment in the training of a woman teacher would, in the long run, be "cheaper" than investing in a man because "a woman will teach for a greater number of years than a man." This, he affirmed, "is because a woman has fewer careers open to her, and thus can devote a greater number of hours to the service of her school." A man, on the other hand, "is always liable to prefer another job, and other business frequently distracts him from providing the school he runs with tireless service."[36]

Covarrubias further commented that "the physiological organization of a woman and her place in society give her a different calling" than the exercise of a profession. However, he saw the need for a woman to cultivate her spirit "with a higher education, which is advisable," since the education of her children was in her hands from the moment they came into the world. It was from mothers, he continued, that children received "their first impressions, first ideas and first knowledge of things," so that one should not leave "a child's first education in the hands of a vulgar intelligence."[37]

After 1867, during the government of Benito Juárez (known as the Restored Republic), the legal reforms that supported the secularization of education were put in place, and a religious morality as replaced by a secular one. This was an important step that allowed public education to expand and more women to enter teaching. In fact, during Juárez's government, women's education was a frequent concern.

As proof we can point to the establishment of the Escuela Secundaria de Niñas, founded in Mexico City in 1869, which granted degrees for primary and secondary school teachers until 1890.[38] In 1889, a decree was issued that transformed the former Secundaria de Niñas into the Escuela Normal para Señoritas. Ignacio Altamirano[39] commented that thanks to this institution, "the future for the poor woman of Mexico would no longer be only the sterile work of a seamstress, or the sad work of a servant, or poverty or something worse. Instead, she could rival a man in certain fields, or surpass him with her greater aptitude in others." He affirmed that society would gain if "there were more educated mothers of families, and more useful women."[40]

In his opening day speech at the Escuela Normal for Señoritas on February 1, 1890, Miguel Serrano[41] said: "I hope that just as the seventeenth century gave men their freedom through education, the nineteenth century will give women equality through education."[42] This school was very popular, and in 1895 registration had to be closed because there was not enough space for all the students who wished to attend.

On the other hand, Justo Sierra told women students: "what you know and what you have learned will not hinder you. That dreadful fear that educated women are badly suited for the home will ultimately die, since an educated and learned woman will be truly fitted for the home, to be a man's companion and share in the education of the family."[43]

During the Porfiriato,[44] diverse ideologues reinforced some of these ideas. Ezequiel A. Chávez,[45] for example, affirmed that primary education was "destined to be given more completely at every moment by women's maternal hands."[46] This social representation of the woman as the ideal teacher, given her maternal talents, was constructed over diverse spaces and times, and served to set feminization in motion. In fact, during the nineteenth century, a pedagogical theory existed that "understood education as an activity that resembled maternity, and was therefore very suitable for women." Accordingly, if "a mother lives for her children, a woman teacher does the same for her students; both are willing to make sacrifices."[47]

Nevertheless, matrimony and motherhood continued to be of utmost importance for women. For that reason, it was only when women did not have the responsibilities of a wife or mother that she could work as a teacher. It was an occupation for single women or widows, as we shall see. When a woman married, it was appropriate that she attend to her home and not work outside. Her care and attention would be directed toward her husband and her own children, not toward other children. Only women teachers who were single or widowed were socially acceptable, as no cultural norms were transgressed.

But how did women experience this feminization of teaching? And how can we learn about their life stories? These two questions are explored in the following section.

AND WOMEN BEGAN TO WRITE

During the colonial era and for much of the nineteenth century, reading was taught first, and then writing came later. Consequently, many individuals who could read did not know how to write, a group that included many nineteenth-century women. It was not until 1883 in the Escuela Modelo, a prototype school directed by Enrique Laubscher in Orizaba, where reading and writing were taught simultaneously for the first time, dispensing with the phonics method that was previously used.[48] This was how the reading–writing method began to spread throughout the Mexican Republic.

Women did not lag far behind men in acquiring writing skills, which allowed them to appropriate one of the "rules" of that era. Writing, and more specifically correspondence, constituted the principal means of communication during the nineteenth century. Juan Luis Vives said that letters were "a conversation between absent people through writing," and had been invented to "transmit to others one's ideas and thoughts, acting in this way as a loyal interpreter and messenger between men."[49] Different authors commented that letter-writing between women could be understood as "a prolongation of the private, of the everyday." Women, they said, appropriated writing in order to write to their husbands or children when they were away, and imbued their letters with their "longings, joys, or sorrows."[50]

The letters we have examined are archived in the Porfirio Díaz Collection. Women teachers used writing to communicate with the President of the Republic and express their needs to him. These are private letters in which women teachers describe the ways in which they experienced the feminization of teaching.

An Encounter with the Teachers of Yesteryear through Writing

The letters written by women teachers in the Porfirio Díaz Collection are quite diverse. Excerpts are presented as an example. They were written between 1908 and 1910, before the Mexican Revolution began. In general, they wrote to Díaz to request something, as they had many needs.

Women Teachers and Their Sorrows

When reading these letters, we can see that women teachers were worried because their pensions had not been approved—one of the most frequent requests. The first of these letters is from Micaela del Valle, who had worked for 35 years, first as a principal in various schools in Puebla and in the state of Mexico, and then as an assistant in the Federal District. She wrote to Díaz that she was not sure that the "blessing of a pension" would fall on her, and ended with the words "I trust you will grant me this favor." In another case, Clementina Sánchez, who had worked for 32 years, requested her pension. However, it was denied because 19 of those years had been spent in schools in Oaxaca, and only 13 in the Federal District as an assistant. The chief of the Department of Primary and Normal School Education, Gregorio Torres Quintero, answered her request with the following: "You have not accrued the time required by the law to be granted a pension." These examples suggest that it was often difficult for many teachers to obtain pensions, as they were only given to teachers who had worked in Mexico City for 30 years.

Many women teachers viewed the approval of their pensions as an act of justice on the part of Porfirio Díaz. They wrote that they were losing their strength, and that they had the right in their old age to be assisted by the "efficient aid of the government." When they finally had to retire without a pension, they wrote once again to the President. One such teacher was María de la Luz Torres, who had gone to live with her daughter and son-in-law. In her letter she commented: "Most Honorable Señor General Don Porfirio Díaz, I am retiring from teaching because I am very tired and sick. I leave the classroom in sorrow because I have not received my pension."[51]

The allusion to a tired, sick body occurs frequently in the correspondence of many of these teachers. In fact, these are women who started teaching when they were young and healthy, and with the passing of time, had aged and become sick.

Much of what is contained in these letters was part of a complicated game of "checks and balances" which in fact allowed the Díaz regime to continue. The Díaz government needed these women teachers in order to educate the Mexican population, and at the same time, these women needed a job that enabled them to survive. Yet the number looking for jobs exceeded the supply. As a result, many teachers were without work and, once again, they wrote to the President.

Eva Torres, who left teaching upon marriage, wrote to Díaz because her husband had died and she could not feed her children, one of whom was sick. Moreover she was two months behind on her rent, and owed 18 pesos. She requested work teaching classes in "hat making or embroidery, or in pattern-making, handicrafts and needlework."

In another letter, Olivia Sánchez, who had also been widowed, begged that she be given a job because due to her poverty she "had to pawn even my teaching certificate." She added, "it is very sad to live on charity when I can still work." Also the teacher Enriqueta Alvarez requested that both she and her daughter be placed in a prefect position, with a salary of 30 pesos per month, so that they could leave the "precarious situation" they found themselves in, with no house, no food, and nothing to wear. The recently widowed Vicenta Corrales asked for "a recommendation for another job in public education," because although her daughter was "old enough to go to school," she could not send her because she did not have enough money to "pay for school expenses." She closed by telling him "what torments me most of all is not being able to educate my daughter."

In general, the writers of all these letters indicated that they had already appealed to the Minister of Public Instruction to "provide an opportunity." On other occasions they were told to "submit their request at the start of the next academic year."[52] These brief examples show that while on the one hand, Mexican statesmen engaged in a certain discourse regarding the feminization of teaching, a different reality existed that is reflected in the needs of these women teachers.

Other letters were sent by teachers who worked in various schools in the capital, asking Díaz for an increased salary. Such was the case for Delfina Rodríguez, who in her letter said that "with hard work and many privations I am able to live and educate my children, but my situation is untenable, because with my low salary I cannot pay for my basic needs." She requested a raise and closed her letter with the words, "I'm a poor teacher who has served the nation for many long years."[53]

The situation for women teachers who lived in rural areas was even more difficult. Angela Pastrana mentions in her letter that the railroad did not even reach the town where she worked. Since the indigenous children did not speak Spanish, she had to teach them the language, and moreover, had to convince parents to send their children to school. She lived far away from her family, and at times had to deal with people who were "backwards, and even hostile." She had put up with this situation for five years, and requested a raise in salary, as the "twenty pesos" she received did not cover her expenses. From Aguililla, Michoacán, Alma del Castillo, the principal of a rural school for girls, wrote to Díaz that she earned "ninety centavos a day," an amount that did not permit her to meet the expenses of living in that town. Soledad C. widow of Rojas also wrote to the President from Ocoyucan, Puebla, requesting financial help because her daughter was sick, and on such a meager salary she could not take care of her.[54]

The social representation of a young, single female teacher often caused problems. This was the case with Rosa Carmona de Gutiérrez, who served in a rural school in Moroleón, Guanajuato, where it was forbidden for married women to work in schools. For this reason she asked Díaz for his help in relocating her family, as her marital status prevented her from working in her profession.[55]

The poor economic situation faced by numerous women teachers made it necessary for them to seek financial help. In some of these letters, such as the one sent by María de Jesús López, they requested "a loan" and promised to pay it back in "installments." López needed it to open a "grocery store," or to buy a "typewriter or sewing machine" in order to earn something extra. In another letter, Elisa Orozco began her petition as follows: "Overwhelmed by the horrors of poverty, I call on you to help me. She affirmed that both the rent on her house and the price of food had increased greatly, and with the little she had saved she was not able to get ahead."[56]

With respect to social background, two spaces can be distinguished: urban and rural. In the cities, social background was quite varied, with teachers from the lower, middle, and even upper classes (who often did not teach). Ezequiel A. Chávez considered that the Normal School was "the best for the intellectual education of young Mexican ladies." He said that at times, some of his students "attended even though they didn't want to become teachers." However, students from "the middle class did want to obtain the rewards granted to teachers of reading and writing."[57] By contrast, in rural spaces we mainly find women from the lower class, for whom teaching was a means to earn a living.

Women Teachers and their Successes

Other teachers, however, did not experience the feminization of teaching with such anguish. We refer to those who were quite successful in their profession, enjoyed a stable salary and were given social recognition for many years. Some traveled to foreign countries, becoming "cultural intermediaries" who upon their return would utilize their knowledge to benefit public education.

Notable among them was Dolores Correa Zapata, who was sent to Berlin, where she had the opportunity to visit a girls' school, a kindergarten, and a Normal School. She later went to Paris and visited girls' schools.

In addition, Raquel Santoyo (director of the primary school annexed to the Normal School, and a member of the Advisory Board of Public Instruction) and Carmen Krauze de Alvarez left the country for the United States, where they perfected their "technical studies for the teaching of manual work." The need to educate young children sent Estefanía Castañeda and Juana Palacios to "model nursery schools" in the United States. With the same objective, Rosaura Zapata was sent first to California, where she observed how "kinders" functioned in the United States, and then later to France, Switzerland, and Germany, to study the "Froebel" system that was used in kindergartens. She later traveled to England, where she visited special schools for mentally retarded children, and open-air schools for "anemic children."

We find another example in Raquel Melgarejo, a language teacher who was sent to France and England, for the purpose of "adding to her knowledge." The principal of the Lerdo de Tejada School, Teresa Guerrero, received a grant from the United States government in order to improve her skills in "school administration." Finally, the teachers Clemencia Ostos and Aurora Gutiérrez thanked Díaz for sending them "to foreign countries on various pedagogical missions."[58]

In another letter, we find Berta Von Glumer, a teacher of young children, who was sent to the United States to improve her knowledge. In that country she studied "the organization and operation of teacher training schools." Von Glumer was notable for both "the creation of kindergartens, and the training of teachers of young children." She later attended Teacher's College at Columbia University and the Froebel Training School for Teachers in New York where she earned a degree in education in 1909. Her grades were so high at the Froebel school that she was named "Honor Student."[59]

Laura Méndez de Cuenca went to Saint Louis, Missouri, which was the first U.S. city to "incorporate kindergarten into the public school system."[60] The experience gained by this teacher during her travels allowed her to place some distance between herself and the official educational system as she compared it to educational practices in other countries. In this way she was able to form her own opinions about methods of learning, curricular content, and textbooks.[61]

The training and prestige of some women teachers qualified them to become principals of important Escuela Normal para Señoritas, such as the one in the Federal District and Puebla. To be a principal, it was necessary to pass an examination, present a letter of good moral standing and have a degree from a Normal School. Principals were also promoted from the ranks of teachers, and there was a period of time during which civil service exams were open for applicants.[62]

One of these principals was Rafaela Suárez, who was born in Colima in 1835. Her father, a craftsman, strove to give his daughter the best possible education, and did not hesitate to hire "special" teachers for his daughters. Rafaela was always noted for her intelligence: her teachers said that she was called to be a "great teacher." At age 13 she enrolled in the school run by Señorita Rubio, where she obtained a professional degree as a "First Class *Preceptora*." However, she wasn't satisfied with this level of knowledge, and took private classes with a noted French teacher, Enrique Matieu de Fossey. In this way, she learned the importance of a "lecture-style education," which she put into practice on a regular basis. Since she did not like to use old textbooks, she took the one or two books she had that were examples of "modern education" and read them aloud to the class. Of the 52 years she worked in education, 28 were spent as principal of the Escuela Normal para Señoritas in Mexico City: from 1890 until her death in 1910. Before that, she was principal of the Secondary School for Girls, where she had worked as assistant director and prefect. She was also a member of the Advisory Board for Public Instruction.[63]

In attempting to learn more about Rafaela Suárez's life, we find signs that lead us to think that she was a simple person. According to sources, she "always dressed in a very modest way" and that in the Normal School in Mexico City, she "always chose the smallest and least congenial space for her room, which she never furnished with anything comfortable, using the smallest pieces of furniture in the building." About Rafaela it was said: "She didn't live for herself, she lived for others."[64]

Like other teachers, she remained single and lived for her work. This was a teacher whose body was neither weak nor sickly, as was the case for many other women at this time, but instead was strong and healthy, resistant to the hardships of a life with many privations. Rafaela Suárez was, in my view, a woman who fulfilled the social representation that had been constructed for teachers during the nineteenth century.

At the Escuela Normal para Señoritas in the state of Puebla, we encounter Paz Montaño, who had been awarded a teacher's degree by the municipal government in Mexico City. She was known as a "very intelligent and deeply cultured person, a poetess," who could also speak English and French very well. In fact, at the Normal School's opening day ceremonies in 1879, it was reported that "señorita Paz Montaño, very gracefully recited several inspired poems." It was emphasized that she had "noble sentiments" and was "a teacher and dear friend" to the students. "Though she was poor, her pocketbook was always open to students who needed help." She served as principal of this Normal School until 1885. When she retired, she opened a private boarding school, and also continued to collaborate with the Normal School as a teacher of literature and grammar.[65]

Curiously, this teacher wrote twelve letters to Porfirio Díaz between 1908 and 1910. In some of them she merely greeted him; in others she told him about her activities as an educator. On one occasion she told him that she would "soon inaugurate the School for Working Women, which will operate in the evenings and be similar to the school in Puebla that was recently opened in January 1910." In another letter she commented that, prompted by the Centennary, she wanted to publish "a small pamphlet which will narrate the main episodes of the War of Independence, and distribute it in schools and to the general population." She also proposed publishing a "Worker's Manual" which would include what was "most essential" for "people of the working class" to know. In one her last letters, she requested a position in the Escuela de Altos Estudios [School of Higher Studies],[66] as she wished to continue her training. Once again we find a woman who never married and dedicated her life to teaching. She was a teacher who even at age 60 (as she declared in her correspondence) still wished to continue her education.

Señorita Paz Montaño was replaced by Juana Ursúa, who was principal of the Normal School from 1885 to 1886. A native of Colima, she had been a student of Rafaela Suárez in her home state. She did not come to Puebla by herself, but was accompanied by her sister Candelaria, who worked as a "Warden." It was said that Juana's religious ideas led her to establish a practice in the boarding school of "praying the Rosary every night and attending Mass on Sundays and holy days." She was also described as a person whose "simplicity was exaggerated" and who was "extremely clean," as one of her duties was the "spotless upkeep of the school."[67] These characteristics of Juana Ursúa allow us to imagine her as a very neat woman, as simple as her teacher Rafaela Suárez. However, her religious beliefs, which extended to her students, did not make her as popular as her predecessor or her former teacher; perhaps that was why she lasted only one year as the principal of that Normal School.

It should be noted that the 12 principals of this Normal School in Puebla were all single women. Some of them had studied at the very same school. There were principals who, according to the descriptions, were worried about "the cleanliness of the school and discipline." Dolores Pérez Muro, a native of Guadalajara, Jalisco, who replaced Juana Ursúa as principal, maintained the religious practices. However, when the principal's job went to former students at the school, the first change was to eliminate these practices and replace them with a "true freedom from religious practices." Thus the teacher Federica Bonilla did not mark Sundays by going to Mass; she began to hold "civic celebrations," especially on "September 15th to commemorate the nation's independence."[68]

We can observe how a generational change brought with it the transformation of certain practices in the everyday life of this Normal School. However, one thing that did not change was the unmarried status of its principals over several generations. Perhaps this is also related to the social representation of the single teacher, who was asked to behave as if she were the mother of all the students. In this way, one can speak of the role they held as "substitute mothers," as Caroline Steedman mentions.[69]

However, this did not mean that all Porfirian women teachers were single; in fact, as seen in some of the above mentioned examples, some were married and had children, and others were widowed.

FINAL REFLECTIONS

To conclude I will reflect on the specific actions of women teachers that constitute part of the feminization of teaching. I refer on the one hand to the women teachers who achieved change, as they slowly moved from a religious morality to a secular one by means of negotiating certain spaces. That was how the women teachers in Puebla from the "new generation" replaced the space of "the Rosary and Mass" by civic celebrations. Curiously, they did not celebrate September 16,—the day of the Call to Independence in Mexico—but rather September 15, which happened to be Porfirio Díaz's saint's day. These spaces were negotiated within an atmosphere of harmonious coexistence, in which many competing ideas were tolerated.

Some teachers also developed certain strategies that allowed them to introduce new ideas, as for example Rafaela Suárez, who got rid of old textbooks and used books that offered a new kind of education in her classes. Undeterred by a limited number of new textbooks, she would read the lessons out loud to the students, as a strategy to "modernize" the education in her school.

On the other hand, Laura Méndez de Cuenca, like many other teachers who wrote to Díaz, appropriated the dominant culture, but at the same time "re-elaborated" it through her valuable opinions, which gave rise to the start of a school culture that was qualitatively different. I refer to the importance of play, gym, and recess time during school hours, which for a large part of the nineteenth century did not exist. Also the changes in school textbooks were notable: foreign textbooks began to be replaced by ones written by Mexican teachers, such as Daniel Delgadillo and Gregorio Torres Quintero, among others.

At the same time, women teachers such as Rosaura Zapata, Berta Von Glumer, Estefanía Castañeda, and Dolores Correa Zapata participated in the educational policy-making of that era, as their training allowed them to contribute their opinions and even bequeath their thoughts to the generations who followed in post-revolutionary Mexico. All this made "powerless" and "invisible" women teachers visible, and they began to obtain a certain degree of power and recognition. Such is the case with Rafaela Suárez, Raquel Santoyo and María Llamas Bello, who were members of the Advisory Board for Public Instruction. Moreover, many women were school inspectors or principals.

As I mentioned, by appropriating writing, women teachers were successful at making themselves be seen. However, they not only appropriated this practice, but also claimed various physical spaces, especially the space of the classroom. There, the woman teacher took over the walls and utilized them to display educational material. At the same time, it was the woman teacher who imposed order and discipline in the room itself as well as in the students. Another important element enters here: discipline, which the woman teacher appropriated as another one of the "rules" of that era.

The documents I have reviewed bring us closer to a certain "face" of teaching, represented by the image of a tired and sick woman who, after having spent many years in the classroom, did not have the right to a pension. Documents describe poor women, often widows, who had to support their families.

Illness, then, becomes an element that appears with frequency, whether in the woman teacher's body or that of a family member. Illness and poverty are linked for these women teachers. Perhaps that is why another element, hygiene, is present, as seen in Juana Ursúa's habits of strict personal hygiene and the cleanliness of her school, or in the ideas of Laura Méndez de Cuenca, who said: "Mexico needs personal hygiene," and called for "soap, a comb, a feather duster, a broom and a water hose." She also inculcated the so-called "modern habits" of "diet and personal grooming." These were young, healthy women who lost their youth and good health in the schools where they served. As the years went by, many of them became sick from performing their job, sometimes because of poor hygienic conditions in their classrooms.

At the same time, these documents lead us to an encounter with a diverse group of women teachers: those who kept teaching despite the fact that their bodies were sick; others, mostly single, who gave their life to teaching despite the low salaries they received. There were also teachers, intelligent and enterprising, who continued to educate themselves, in order to make innovations and advance the cause of public education. And at the same time there were teachers who were not satisfied with their level of knowledge, and who studied languages or applied for entrance to the Escuela de Altos Estudios.[70]

This reality contrasts with another social representation constructed by Ezequiel A. Chávez when he wrote: "those delicate women who go to distant villages . . . to create with the efficiency of their word and the allure of their kindness the light of science." Indeed, these women teachers traveled to the furthest corners to transmit the knowledge of their era, as shown by the letters they wrote. Their bodies were not fragile; on the contrary, they were strong enough to endure illness, poverty, and at times even the rejection of the community they served.

Finally, we also encounter social representations that were constructed around the figure of the woman teacher as an unmarried woman. Perhaps it was for this reason that she was rejected by society when she married, because it had equated her with the figure of the mother. Unconsciously, she was asked to be a "loving mother" to all of her students, and in order to carry out this role she had to be free, a single woman by necessity. To this social representation was added the one that regarded women as the best fit to teach, which could be discerned in several public speeches. Through these representations the feminization of teaching was set in motion. To work as a teacher did not "masculinize" a woman; instead, it was part of her feminine identity.

Through letters which speak to us of certain social representations, we have been able to get closer to the women teachers of yesteryear, the educators who offered their services during the Porfiriato. This is, however, only a preliminary reading of these source materials. Many letters, photographs, and other documents are waiting to be examined, and further readings remain to be done.

NOTES

1. Porfirio Díaz occupied the Presidency for 30 years, with an interval from 1880–1884 during which his close friend Manuel González served as President. The Constitution was then amended to allow the President to be re-elected. It was not until November 20, 1910, when Francisco Madero led an armed insurrection in the state of Chihuahua, that the dictatorship began to crumble. Díaz left the country in May 1911 and died abroad. During his government the economy improved: foreign capital was invested, the industrialization of the country was initiated and railroads began to be constructed, mainly toward the north of the Republic as the United States was the principal destination for exports. However, while a small segment of society became wealthy, the majority of Mexicans lived in great poverty. As a result, when Francisco Madero made his call to arms in 1910, the movement was very successful. Over the course of the following decade, different political bosses emerged, and the revolutionary movement ended in 1921.
2. Jane Dale Lloyd, "Preliminares," *Historia y Grafía* (México: Universidad Iberoamericana, Núm.13, 1999), p. 8
3. The long period of Porfirio Díaz's government is known as the "Porfiriato."
4. Jane Dale Lloyd, "Preliminares," p. 10.
5. Ibid., p. 11.
6. In Mexico, both the Ministry of State and the Office of Justice and Public Instruction (1862–1905), as well as the Ministry of Public Instruction and Fine Arts (1905–1917), operated only within the

Federal District and the Territories (Quintana Roo, Tepic, and Baja California). It was not until 1921 with the establishment of the Ministry of Public Education that the educational system reached national coverage.

7. The term "*preceptora/preceptor*" was used throughout the nineteenth century to identify women and men teachers. The word derives from the Latin "*praeceptor*" meaning teacher. It initially referred to an educator who instructed children at home. Later, the use was extended to teachers who worked in public or private schools. Primary school *preceptores* could be of first, second, or third class, depending on the number of subjects they had studied. First-class *preceptores* studied the greatest number of subjects. (Luz Elena Galván (ed.), *Diccionario de Historia de la Educación en México*, CONACYT, CIESAS, DGSCA/UNAM, 2002. Sección de términos).

8. Ibid.

9. Gabino Barreda was the founder of the Escuela Nacional Preparatoria, considered to be the most important educational project during the administration of Benito Juárez (1858–1867).

10. Mary Kay Vaughan, *Estado, clases sociales y educación en México* (México: SEP/80 and FCE, 1982), p. 39.

11. Abelardo Villegas, *Positivismo y porfirismo* (México: Sep-Setentas, 1972), p. 75.

12. Francisco Cosmes, *Historia General de México*, ed. Ramón de S.M. Araluca, (Mexico: 1901), p. 268.

13. Leopoldo Zea, *Del liberalismo a la revolución en la educación mexicana* (México: Instituto Nacional de Estudios Históricos de la Revolución Mexicana, 1956), p. 113.

14. Manuel Dublán and José María Lozano, *Legislación Mexicana o Colección completa de las disposiciones legislativas expedidas desde la independencia de la República* (México: Imprenta del Comercio de Dublán y Chávez, 1910), vol. 10, p. 582.

15. Francisco Cosmes, *Historia General de México*, pp. 128 and 129.

16. Ibid., p. 915.

17. The system of objective education proposed that different objects and their names be taught at the same time. In Mexico this method was introduced by Manuel Guillé and Vicente H. Alcaraz (Ernesto Meneses et al., *Tendencias educativas oficiales en México, 1821–1911* (Mexico: Porrúa, 1983), pp. 568 and 569).

18. Ernesto Meneses et al., *Tendencias educativas oficiales en México*, pp. 350 and 354.

19. For an example of how these ideas arrived in the state of San Luis Potosí, see: Adriana Mata Puente, *La escuela y la lectura en San Luis Potosí durante la segunda mitad del siglo XIX*, (master's thesis, Colegio de San Luis, Mexico, April 2003).

20. Quintana Roo, Tepic and Baja California.

21. Manuel Dublán and José María Lozano, *Legislación Mexicana,* vol. 19, p. 127.

22. Mary Kay Vaughan, *Estado, clases sociales y educación en México,* pp. 40 and 41.

23. Justo Sierra was the Minister of Public Instruction during the final years of the Porfiriato (1905–1911).

24. Justo Sierra, *La educación nacional. Artículos, actuaciones y documentos,* Edited and annotated by Agustín Yáñez (México: UNAM, 1948), pp. 103, 110, and 185 (Obras completas del Maestro Justo Sierra, vol. 8).

25. Joaquin Baranda was Minister of Public Instruction during the Porfiriato from 1884 to 1901.

26. Ezequiel A. Chávez, "La educación nacional," in Justo Sierra, *México, su evolución social,* vol. 1 ed. J. Ballescá y Compañía Sucesor (Mexico: 1902), p. 556.

27. Mary Kay Vaughan, *Estado, clases sociales y educación en México,* pp. 41–42.

28. Ezequiel A. Chávez, "La educación nacional," pp. 567 and 571.

29. José María Puig Casauranc, *La educación pública en México a través de los mensajes presidenciales desde la consumación de la Independencia hasta nuestros días,* (México: SEP, 1926), p. 124, cited in Luz Elena Galván, *Los maestros y la educación pública en México* (México, CIESAS, 1985), pp. 27–28.

30. Toward 1908, Sierra reiterated to Limantour that the education budget was essential, but Limantour was not persuaded. He transferred out of the Treasury 450,000 pesos that were earmarked for public education, thus eliminating salary increases to teachers.

31. Secretaría de Economía, México, 1956, p. 239, cited in Luz Elena Galván, 1985, *Los maestros y la educación pública en México,* p. 30.

32. See Rudé, *La multitud en la historia* (México: Siglo XXI editores, 1979).

33. See Luz Elena Galván, "Del arte de ser maestra y maestro a su profesionalización," in Luz Elena Galván (ed.), *Diccionario de Historia de la Educación en México,* Multimedia version, Mexico, CONACYT, CIESAS, DGSCA/UNAM, 2002.

34. The concept of "easy access" means that teacher trainees did not have to undertake long years of study as did, for example, laywers, engineers, and doctors. Girls began their training at a very young age, at 11 or 12, and after four years of study, began working as teachers at age 15 or 16. See San Román, *Las primeras maestras. Los orígenes del proceso de feminización docente en España* (Barcelona: Ariel, 1998).

35. José Díaz Covarrubias, *La instrucción pública en México. Estado que guardan la instrucción primaria y la secundaria y la profesional en la República. Progresos realizados. Mejoras que deben introducirse* (México: Imprenta del Gobierno en Palacio, 1875), cxx.

36. Ibid., cxxii.

37. Ibid., clxxxvii and cxciii.

38. With respect to the definition of a lay teacher, four categories are possible: "1) is independent from the jurisdiction of the ecclesiastical authorities; 2) abstains from all forms of religious instruction; 3) does not admit any clergy or minister into the instruction; 4) is not related to religious associations or corporations" (Meneses et al., *Tendencias educativas oficiales en México*, p. 718). Also consult the article by Norma Durán, "La invención del laico en la Edad Media," *Historia y Grafía* (México: Universidad Iberoamericana, no. 16, 2001). Regarding the establishment of secondary schools for girls, see Ezequiel A. Chávez, "La educación nacional," in Justo Sierra, *México: Su evolución social*, (1948), p. 548.

39. Noted writer and teacher who advocated a national-oriented curriculum.

40. Ignacio Manuel Altamirano, *El Renacimiento. Periódico Literario (México 1869)*, "Introduction," by Humberto Bátiz (México: UNAM, 1979), p. 388.

41. Serrano was the director of the Escuela Normal para Profesores, an all-male teacher training school.

42. Moisés González Navarro, "El Porfiriato. La vida social," in Daniel Cosío Villegas, *Historia Moderna de México* (México: Editorial Hermes, 1973), p. 666.

43. Justo Sierra, *México: Su evolución social*, p. 329.

44. The long Presidency of Porfirio Díaz (1876–1910) is known as the Porfiriato.

45. Ezequiel A. Chávez was the Deputy Minister of Public Instruction under Justo Sierra (1905–1911).

46. José Díaz Covarrubias, *La instrucción pública en México*, p. 562.

47. Gabriela Cano, "Género y construcción cultural de las profesiones en el porfiriato: magisterio, medicina, jurisprudencia y odontología," *Historia y Grafía* (México: Universidad Iberoamericana, no. 14, 2000), p. 207.

48. Mílada Bazant, *Historia de la educación durante el porfiriato* (México: El Colegio de México, 1993), p. 53.

49. Juan Luis Vives, *Epistolario*, Edición de José Jiménez Delgado (Madrid: Editora Nacional, 1978), p. 599.

50. See works by Antonio Viñao Frago, *Leer y escribir. Historia de dos prácticas culturales* (Mexico: Fundación Educación, voces y vuelos, I.A.P., 1999) and Antonio Castillo Gómez, *Escrituras y escribientes. Prácticas de la cultura escrita en una ciudad del renacimiento* (Gobierno de Canarias: Fundación de Enseñanza Superior a Distancia de Las Palmas de Gran Canaria, 1997).

51. *Colección Porfirio Díaz* at the Universidad Iberoamericana. Legajo 33, Caja 6, Documentos: 2338 and 2339; Caja 8, Documento 3011, 3101 and 3102; and Caja 15, Documento 5643. [In Mexican archival systems, "Legajo" means Folder; "Caja" means Box; "Documento" means Document. Trans.] This Collection archives the documents

that were received daily by President Porfirio Díaz between 1876 and 1915, the year he died in exile. Those wishing to learn more about this collection can consult Luz Elena Galván, *Soledad compartida. Una historia de maestros, 1908–1910* (México: CIESAS, 1991), in which I reviewed more than 60,000 letters written by a variety of Mexicans between 1908 and 1910, in order to identify those written to the President by teachers. For this article I selected letters that were written by women teachers.

52. *Colección Porfirio Díaz*, Legajo 33, Caja 3, Documento 851; Caja 4, Documento 1410; Caja 10, Documentos 3620 and 3621; and Legajo 34, Caja 7, Documento 3407.

53. Ibid., Legajo 35, Caja 1, Documento 497.

54. Ibid., Legajo 34, Caja 35, Documento 17180; Legajo 33, Caja 6, Documento 2064; and Caja 15, Documento 5849.

55. Ibid., Legajo 34, Caja 38, Documento 18944.

56. Ibid., Legajo 33, Caja 9, Documento 3454; and Caja 4, Documento 1527.

57. Ezequiel A. Chávez, *"La educación nacional,"* 562.

58. Archivo General de la Nación, Fondo Instrucción Pública y Bellas Artes, Caja 230, Expediente 6 and 7. Colección Porfirio Díaz at the Universidad Iberoamericana, Legajo 33, Caja 12, Documento 4551; Caja 23, Documentos 4080 and 9035; Caja 39, Documento 15226; Legajo 34, Caja 35, Documento 17186; and Legajo 35, Caja 8, Documento 3781. Also see Rosaura Zapata, *La educación preescolar en México* (Mexico: n.p., 1951), pp. 11 to 25.

59. Alejandra Zúñiga, "Berta von Glumer: Su vida académica en el nivel preescolar (Biografía)," Manuscript, 2001, pp. 3–4.

60. Mílada Bazant, "Laura Méndez de Cuenca y su visión educativa del porfiriato," Paper presented at the First International Congress on the Process of the Feminization of Teaching, San Luis Potosí, Mexico, February 21–23, 2001, p. 4.

61. Ibid., pp. 4–12.

62. *Colección Porfirio Díaz*, Legajo 33, Caja 3, Documentos 3606 to 3609, Caja 23, Documento, 11130, Legajo 35, Caja 10, Documento 4741, and Caja 18, Documento 8669.

63. The Advisory Board for Public Instruction [El Consejo Superior de Instrucción Pública], was created on August 30, 1902. It was founded to maintain "harmony and coordination among the institutions that serve the State, and to promote the progress of future generations." (Archivo Histórico de la SEP, 1929, expediente 16, cited in Luz Elena Galván, *Los maestros y la educación pública en México*, p. 28.

64. Francisco Hernández Espinosa, *Historia de la Educación en el Estado de Colima* (México: Gobierno del Estado de Colima, 1961), pp. 78–85.

65. *SURSUM*, Revista Mensual, Puebla, no. 17 and 18, September 1955: 14, 15, 24, and 25; and Rubén Gracia, *La Escuela Normal de Puebla* (Puebla: Gobierno del Estado de Puebla, 1950), p. 28.

66. *Colección Porfirio Díaz*, Legajo 35, Caja 4, Documento 1719, Caja 8, Documento 3738.
67. *SURSUM*, Revista Mensual, footnote 29, p. 25.
68. Ibid., pp. 25–26.
69. Caroline Steedman, "The Mother Made Conscious: The Development of a Primary Pedagogy," *History Workshop Journal*, 20, 1985.
70. This school, founded by Justo Sierra in 1910, was part of the Universidad Nacional de México, and was the forerunner to the Escuela Normal Superior.

ARCHIVAL SOURCES

Archivo General de la Nación. Ramo Instrucción Pública y Bellas Artes.
Colección Porfirio Díaz. Materials relating to the years 1908–1910. Locations cited in the text as Legajo [Folder], Caja [Box], and Documento (Document).
Collection "Genaro García" in the Nettie Lee Benson Latin American Collection. Library of the University of Texas, Austin, U.S.A.

Notes on Contributors

Sandra Acker is a professor in the Department of Sociology and Equity Studies, Ontario Institute for Studies in Education of the University of Toronto, and associate dean (Social Sciences), School of Graduate Studies, University of Toronto. She has worked in the United States, Britain, and Canada. Her general field is sociology of education, with special interest in gender and education, teachers' work, and higher education. Recent book publications include: *The Realities of Teachers' Work: Never a Dull Moment* (Cassell, now Continuum, 1999) and co-edited with Elizabeth Smyth and others, *Challenging Professions: Historical and Contemporary Perspectives on Women's Professional Work* (University of Toronto Press, 1999).

Juan Alfonseca has a master's degree in Social Science from the Facultad Latinoamericana de Ciencias Sociales in Mexico City. He is currently a doctoral student in Latin American Studies in the Department of Political and Social Sciences at the Universidad Nacional Autónoma de México. His significant publications are: "Escuela y sociedad en los distritos de Texcoco y Chalco," in Luz Elena Galván (ed.), *Miradas en torno a la educación de ayer* (México: COMIE, 1997); "La escuela socialista en la región de los lagos y los volcanes," in Alicia Civera (ed.), *Experiencias educativas en el Estado de México. Un recorrido histórico* (Toluca, México: El Colegio Mexiquense, 1999); "La escuela rural federal y la transformación de los supuestos sociales de la escolarización en una región del Estado de México. Los Distritos de Texcoco y Chalco 1923–1940," in Luz Elena Galván et al. (eds.), *Historiografía de la Educación en México* (México: Consejo Mexicano de Investigación Educativa A.C., 2003).

Regina Cortina is an associate professor in the School of Education at the University of North Carolina in Chapel Hill. She has a Ph.D. in Social Sciences in Education from Stanford University's School of Education. Her research focuses on the training and employment of teachers and on issues of equity and politics in education. In addition to her book *Immigrants and Schooling: Mexicans in New York* (New York: Center for Migration Studies, 2003), she has recently published two books related to issues of education in Latin America: *Líderes y construcción de poder: Las maestras y el SNTE* (México: Santillana, 2003), and *Distant Alliances: Promoting Girls' and Women's Education in Latin America* (New York: Routledge, 2000). At present, Professor Cortina is conducting research on migration and education focusing on how schools can better engage the educational needs of the growing Latin American-born population in the United States.

Marc Depaepe is a doctor of Social Sciences in Education. He is full professor of History of Education at the Katholieke Universiteit Leuven in Belgium. He was chair from 1991 to 1994 of the International Standing Conference for the History of Education (ISCHE) and a member of the editorial advisory board for the journal *Paedagogica Historica*. Professor Depaepe has edited and published approximately 300 books and articles in 12 languages on the history of education in Belgium, the methodology and theory in the history of education, and experimental pedagogical sciences. His most important works are: *On the Relation between Theory and History of Education. An Introduction to the German Discussion on the Relevance of the History of Education* (Leuven: 1983); and *Order in Progress. Everyday Educational Practices in the Primary School (Belgium: 1880–1970)*, (Leuven: 2000).

Luz Elena Galván Lafarga has a Ph.D in History from the Universidad Iberoamericana in Mexico and has been a researcher at the Centro de Investigaciones y Estudios Superiores en Antropología Social (CIESAS) since 1974. Her research interests are the history of education in the nineteenth and twentieth century, and the teaching of history. She is a member of the National System of Researchers Level II and the Mexican Council of Educational Research. Her most important publications include: *La educación superior de la mujer en México, 1876–1940* (México: CIESAS, 1984); *Soledad compartida. Una historia de maestros (1908–1910)* (México: CIESAS, 1991); editor of *Miradas en torno a la educación de ayer* (Guadalajara: COMIE-Universidad de Guadalajara, 1997); editor of the *Diccionario de Historia de la Educación en México. Versión Multimedia.* Second edition (México: CONACYT, 2003); and editor of *Historiografía de la Educación en México* (México: COMIE, 2003).

Hilde Lauwers has an MA in modern history (Vrije Universiteit Brussel). During several years, she was a scholar in a project between the K.U.Leuven, the Universiteit Gent, and the teachers' union COV, which investigated the history of primary education in Belgium. Among other works, she collaborated on the publication of Marc Depaepe, *Order in Progress. Everyday Educational Practices in the Primary School (Belgium, 1880–1970)* (Leuven: 2000) and has focused on the feminization of the teachers' corps in Belgium (1945–2000) and more generally on gender and education. Currently, she works for the Research Center Child and Society (Belgium).

Iván Molina works at the School of History and the Center for Research on Latin American Identity and Culture (CIICLA) at the Universidad de Costa Rica. Some of his publications are: *Costa Rica (1800–1850). El legado colonial y la génesis del capitalismo* (San José: Universidad de Costa Rica, 1991); *Héroes al gusto y libros de moda. Sociedad y cambio cultural en Costa Rica (1750–1900)* (San José: Ed. Porvenir y Plumsock Mesoamerican Studies,

1992) co-edited with Steven Palmer; *Urnas de lo inesperado. Fraude electoral y lucha política en Costa Rica (1901–1948)* (San José: Universidad de Costa Rica, 1999) written with Fabrice Lehoucq; *Educando a Costa Rica. Alfabetización popular, formación docente y género (1880–1950)* (San José: Plumsock Mesoamerican Studies y Ed. Porvenir, 2000) written with Steven Palmer; *La ciudad de los monos. Roberto Brenes Mesén, los católicos heredianos y el conflicto de 1907 en Costa Rica* (Heredia: Editorial Universidad Nacional y Universidad de Costa Rica, 2001); and *Costarricense por dicha. Identidad nacional y cambio cultural en Costa Rica durante los siglos XIX y XX* (San José: Universidad de Costa Rica, 2002).

Graciela Morgade has a master's degree in Social Sciences from the Latin American School of Social Sciences, in Buenos Aires. She received her bachelor's degree in Educational Science from the Universidad de Buenos Aires. Among her more important publications are the books: *Mujeres en la educación* (Buenos Aires: Miño y Dávila Eds., 1997) ed.; *Aprender a ser mujer, aprender a ser varón* (Buenos Aires Ediciones Novedades Educativas, 2001); *El determinante de género en el trabajo docente de la escuela primaria* (Buenos Aires: Miño y Dávila Eds., 1995). The articles she has published include "Tecnologías de género y carrera profesional docente: desafíos de las mujeres en un sistema educativo feminizador," in *Revista Nómades* of the Universidad Central de Bogotá, April 2001, No. 14; "Dinámicas de género en los discursos constitutivos de la gestión de las escuelas primarias argentinas," *Educaçao & Realidade*, Universidad Federal de Rio Grande do Sul, No. 25.

Elisabeth Richards is currently teaching at the Ontario Institute for Studies in Education of the University of Toronto. Her graduate studies have been in the areas of education and sociology and she has taught both in North America and in Asia. Her research interests include first and second language acquisition, the sociology of work, qualitative research, higher education, and marginal subjectivities. Her most recent publication is "Finding their way: An exploration of immigrant students in an inner city in Toronto," (with Grace Feuerverger), in International Handbook of Diversity (in press).

Fúlvia Rosemberg is Brazilian. She holds a doctorate in Psychology from the University of Paris. Currently, she is a researcher at the Fundação Carlos Chagas, coordinating the Brazilian site of the Ford Foundation International Fellowship Program; she is also a graduate professor at the Universidad Católica de São Paulo. Her research focuses on gender hierarchies, race, and gender in education. Some of her recent publications include "Gender Subordination and Literacy in Brazil," "Education, Democratization and Inequality in Brazil," "Education, Inequality, and Race in Brazil," and "Ambiguities in Compensatory Policies: A Case Study from Brazil," in Regina Cortina and Nelly P. Stromquist (eds.), *Distant Alliances: Promoting Girls' and Women's Education in Latin America* (New York: Routledge, 2000).

Sonsoles San Román obtained a Ph.D. in Sociology at the Universidad de Salamanca with a Doctoral Thesis on the "Social Transformations of the Profession of Women Schoolteachers." Since 1987, Dr. San Román is a professor in the Department of Sociology at the Universidad Autónoma de Madrid, where she has also participated in the Department of the Philosophy of Education. In addition, she has been a visiting scholar at the Institute for Education at the University of London. Among Dr. San Román's recent publications are the following titles: *Las primeras maestras. Los orígenes del proceso de feminización docente en España* (Barcelona: Ariel, 1998); *Fases de incorporación de la maestra a la escuela pública en España (1873–1882)* (Colegio de San Luis Potosí, San Luis Potosí, México: 2001; and *Espacios históricos generacionales de la maestra en el proceso de cambio social de transición democrática en España* (Madrid: Instituto de la Mujer, Ministerio de Asuntos Sociales, 2002).

Frank Simon obtained his doctorate in History at the Universiteit Gent, Belgium. He is a professor of the History of Education at the Universiteit Gent and was a professor at the University of Brussels. His research interest is elementary education in the nineteenth and twentieth centuries. Professor Simon is editor of the journal *Paedagogica Historica, International Journal of the History of Education,* and a member of the editorial advisory board of the *Anuario de la Sociedad Mexicana de Historia de la Educación.* He was also on the editorial advisory board of the journal *History of Education* and a member of the board of directors of the History of Education Society in the United States.

INDEX

Printed in the United States
By Bookmasters